Shaping the College Experience Outside the Classroom

Shaping the College Experience Outside the Classroom

by
James Scannell

and
Kathleen Simpson

The University of Rochester Press

First published 1996

University of Rochester Press
34–36 Administration Building, University of Rochester
Rochester, New York, 14627, USA
and at P.O. Box 9, Woodbridge, Suffolk IP12 3DF, UK

ISBN 1–878822–68–3 (Hardback)

Library of Congress Cataloging-in-Publication Data

Shaping the college experience outside the classroom / by
James Scannell and Kathleen Simpson.
p. cm.
Includes bibliographical references and index.
ISBN 1–878822–68–3 (alk. paper)
1. Internship programs—United States. 2. Practicums—United States.
3. College students—Employment—United States.
4. College student development programs—United States.
I. Simpson, Kathleen. II. Title.
LC1072.I58S33 1996
378.1′78—dc20
95-51462

British Library Cataloguing-in-Publication Data

A catalogue record for this book
is available from the British Library

This publication is printd on acid-free paper
Printed in the United States of America

This book is dedicated to Kathy Kurz:

. . . valued colleague

. . . special friend

Contents

ACKNOWLEDGMENTS

Like most, if not all, evolutionary (and if one is lucky revolutionary) ideas, the programs, activities, and outcomes studied on the following pages are the result of the energy, wisdom, dedication, and even intervention of many, many people.

I would like to recognize my colleagues, Caroline Nisbet at Duke University and Emily Newton at the University of Rochester, whom I had the privilege to work alongside in Ithaca during the creation and implementation of the idea, concepts, and founding principles of The Cornell Tradition. I also want to acknowledge the late Robert Kreidler who was President of the Charles A. Dana Foundation as well as Stephen Barkanic and D.J. Jensen, Program Officers at the Foundation, for their interest, support, thoughtful advice, and counsel. I want to thank all of the staff and faculty and especially the alumni of these programs at Cornell, Duke, the 20 participating Dana Foundation institutions, and Rochester for their commitment and loyalty to the financial support of students in search of a college education. This very special commitment helps pave the way for a civic-minded citizenry and the continuation of our institutions of higher education. I thank all of you for your help, guidance, and cooperation in this effort.

Over the course of this three year research project, a number of students have been involved, most appropriately given the subject matter, in data collection, coding, data entry, communication, and follow up: Larry Boodin, Courtney Clark, Sarah Collard, Tina Dankos, Frances Fiocchi, Apurv Garg, Irene Maiolo, Bridget Munn, Susan Putnam, Stacey Spall, and Jeanne Tetreault. I've also received wonderful research support from Jim Henson of the University of Rochester, Michelle Tullier of Barnard College, and Michelle Bailey of Maguire Associates. I have also had the benefit of three excellent reviewers, Richard Skelton, Janiece Bacon Oblak, and Anthony Lolli. Most especially I want to express my gratitude and appreciation to Michael Posner, who took on the task of research assistant. Rita Haschmann, my longstanding assistant who patiently and unselfishly produced the manuscript, maintained the balance.

Jim Scannell
Rochester, NY
March, 1996

To those noted by Jim Scannell already, I add my sincerest gratitude. I especially wish to thank Emily Newton, experiential guru, who introduced me to my first Reach experience during my student days. I also want to acknowledge Pearl Rubin, former President of the Daisy Marquis Jones Foundation, who encouraged my involvement in program evaluation. For sharing their historical roots of experiential learning and creative partnerships with industry, I owe tremendous thanks to folks at Berea College in Kentucky and Barry College in Georgia who graciously and generously gave their time during my visits there. Wayne Locust, my director, deserves special thanks for his support and encouragement. And because work on the book coincided with other full-time responsibilities, I owe a tremendous debt of gratitude to my husband Gary and children, Kirk, Scott, Thai and Dan, for their unflagging support and good humor.

Kathy Simpson
Rochester, NY
March, 1996

Prologue

Once upon a time, in the not too distant past, it was possible to aggressively pursue two or maybe even three summer jobs (some full time, some part time), work 20 hours a week during the academic year, and, provided a certain level of frugality, be able to produce a bottom line of savings that met most of the annual expense of going to college. This was true in spite of the fact that wages were much less, because of course so were college tuitions. Thirty years ago, for instance, even at private institutions, tuition was less than $2000 a year, the average being closer to $1000. Sad to say that those days are past and are not likely to return.

Cornell developed a program to recognize the spirit of working your way through college, the commitment to the work ethic, the ownership of responsibility, and all of the attendant skills and values associated with this effort. President Frank Rhodes repeatedly had raised the pressing issues of financing a Cornell education. The program Cornell created challenged the popular opinion of the day, namely that one was entitled to a college education, while at the same time responded to concern about levels of debt students were taking on and what that would mean to their futures in terms of career choice, graduate education, or the ability and willingness to contribute as citizens to society, volunteering their time and financially supporting their institution for the good of the whole.

With these objectives a set of programs emerged, all of which focused on the creation of educationally purposeful, career-related, paid employment opportunities that would both help students pay for a significant fraction (perhaps a quarter or third of their college costs) and/or reduce their indebtedness. At the same time these programs provided opportunities to further develop plans, aspirations, and dreams whether in academic or professional careers. These programs, known as The Cornell Tradition, were announced in the fall of 1982 and became operational with the Summer Job Network and Summer Fellowship Program of The Cornell Tradition in June, 1983.

One year after its inception, Jim Scannell was asked to make a presentation in New York City to a gathering of foundation officers interested in discussing the challenges and issues students and families, as well as institutions, faced in meeting the cost of college. His presentation focused on the anticipated demographic decline, disinterest on the part of the Reagan Administration in funding federal student aid, and the need for colleges and universities to control costs. The presentation ended with an introduction and explanation of The Cornell Tradition program. At lunch Robert Kreidler, then President of the Dana Foundation, sought Jim out. He wanted to hear more. From that brief discussion, a series of meetings were held over the course of the next year, the product of which was an aggressive, precedent-setting initiative by the Foundation to seek proposals, mostly from liberal arts colleges, for the development of creative, financial aid, and student work programs known as the Dana Student Aid for Educational Quality (SAEQ) Program. Colleges participated by invitation only. Institutions were challenged to develop proposals that met most if not all of the following goals and requirements: enhanced student recruitment, an experience of high educational value, increased faculty productivity and involvement, debt reduction, increased participation of alumni, new corporate support (financial or employment opportunities), a sound plan to evaluate the success of the program in meeting the stated goals, a stated commitment to secure matching funds (two to one was the going rate), the likelihood of the program continuing at their institution, and the probability of the program being replicated at other colleges and universities. Over the course of the next three years, more than 60 institutions submitted proposals, 28 of whom each received $200,000 Dana Foundation Student Aid for Educational Quality grants to cover a four-year period of program activity. Steve Barkanic, Program Officer at Dana at the time, staffed the proposal review process. Jim had the privilege of joining a review team consisting of Rex Moon, retired College Board official, and Aaron Liminick, who at that time was Provost at Princeton University. 1985 launched the beginning of the Dana SAEQ programs. The first round of winners included such schools as Bates College, Bryn Mawr, Bucknell University, Dickinson, Furman University, and Oberlin. In the subsequent three years additional institutions were added to this original group. A total of 28 were funded over the life of the program.

Also in 1985 Duke University launched an internship program called Duke Futures. Caroline Nisbet, who had been Director of Student Employment at Cornell University and one of the architects of The Cornell Tradition, developed the Duke Futures program, creating yet another variation on the theme of The Cornell Tradition. Finally, in 1986, Emily Newton, Assistant Director of Student Employment at Cornell and first coordinator of The Cornell Tradition, joined the Enrollments, Placement, and Alumni Affairs division of the University of Rochester. Shortly thereafter the Rochester version of The Cornell Tradition, Reach for Rochester, was launched. So much for history.

This book is about whether it worked. Whether these programs met their established goals. And, just how was it done? Whether foundation program officer, benefactor, student development specialist, or college president, it is assumed that those involved in creating and implementing educational programs want to be confident that their efforts and support are as effective and efficient as possible; that the right partners are being brought into the play; that the program in fact is making a difference. Educationally purposeful, career-related, paid employment opportunities such as those described above all have the characteristics necessary to meet these high expectations. This book will document the outcomes.

The programs cited earlier—The Cornell Tradition, Duke Futures, the Dana Foundation Student Aid for Educational Quality, and Reach for Rochester—have all demonstrated the ability to meet multiple goals and objectives with high style including:

1. The development of individual self worth and a confidence due to the direct reward for the work ethic. The value of earning power, when part of the funds necessary for a college education is earned through work, is duly recognized.
2. The development of successful recruitment strategies for institutions desiring to attract the best academic, socio/economic, and racial ethnic mix of students by offering opportunities to gain hands-on experience in fields that will contribute significantly to career exploration. These strategies boast added value to the collegiate experience.
3. The creation of meaningful connections with prior generations of students (alumni/ae), the development of a sense of obligation in the present generation to maintain ongoing

financial support and career opportunity for future generations of college students.

4. The opportunity to efficiently engage alumni in meaningful activity to which even young alumni who have yet to achieve financial security can make a significant contribution on behalf of their institution for the next generation of undergraduates.
5. The chance to engage employers and let them sample the quality of an institution's product, its currently enrolled students. This introduction to campus community has proven to be a tried and true strategy for the recruitment of those students to first-career, full-time positions after graduation.
6. The opportunity to involve the academic enterprise (faculty) through undergraduate research assistantships while simultaneously advancing the faculty's scholarly work. Students were introduced to the world of the Academy and came to understand what faculty do, all of which is critical if our society is to fulfill its professorial needs in the future. At the same time, faculty had the opportunity to interact with students on a different level, in a different environment, in a different relationship—a win-win experience.
7. The chance for institutions to join with the community at large to address large scale social, economic, environmental problems and challenges. Service is the third goal of institutions of higher learning after teaching and research.
8. The opportunity to leverage donated funds in support of these programs:
   a. By creating a new career-related job through the subsidization of the gross wage. The support to the student was, for example, leveraged three to one (if the subsidy is 33%) for a position that would not have existed otherwise.
   b. By creating endowed gift opportunities for special fellowships or scholarships which reduced indebtedness because of the value placed on public service and/or work. These debt-reducing awards in recognition of work allowed students to put a real dent in educational costs when combined with the hourly wage earned.
   c. By establishing matching opportunities for foundations and corporations eager to support "self-help" programs.
   d. By promoting challenge grants from individuals and re-

union classes (e.g., from older, established classes to younger classes of recent graduates).

e. By using revolving funds as seed money to develop student enterprises with the potential for profit to the institution, which could in turn be used to support additional student efforts.

f. By merging private funds, federal dollars, and institutional resources—an ideal partnership (e.g., FIPSE community service grants and institutional funds).

9. The opportunity to create a synergy where the sum of all of the above is greater than the individual parts. This opportunity has the potential to become part of the institution's signature or the signature itself. These opportunities also have the potential to make and shape careers and lives.

An initial draft of the manuscript and research findings was reviewed in the fall of 1994 at a meeting entitled "*Conference on Career-Related Work and Student Aid*" in Rochester, New York. Over 40 participants from Cornell, Duke, the Dana schools, and the University of Rochester attended the two days of discussion. The purpose of the conference was two-fold: to review the manuscript and to share ideas.

One outcome of the conference was a crisper definition of the audience for whom this book is written, that is, in general, student services and student development staff at colleges and universities. This group broadly defined would include: career planning and student employment staff; admissions and financial aid officers; those whose duties include enrollment management/planning; deans of students and student affairs officials; community college transfer and placement officers; and high school guidance counselors. Namely, those in the front line providing direct service to students. The goal is to provide those who track, shape, and help students develop their lives with ideas, programs, activities, and projects that will help them do their jobs, help them develop or realign programs. It is meant to encourage and hopefully inspire. That is why this is not a scientific, psychometric, personal development study of individuals but rather a review of programs that share similar concepts and goals and an analysis of the impact those programs have had on the lives of individuals and the development of institutions.

In addition, attendees at the conference stated a belief that there is a large and very broadly defined secondary audience for this story, namely faculty, institutional advancement officers, and senior level collegiate and university administrators. Faculty will find the use of undergraduates in research and the match of experiential task to educational outcomes of interest. Fundraisers will focus on the attractiveness of this "product line" to potential donors both individuals as well as corporations and foundations. Alumni officers of course are always in search of new linkages with their constituents. And, finally, administrators in search of institutional distinctives, valued programs, and cost benefit will find the experiences of both individuals and institutions instructive.

In sum, this book attempts to identify what works and what doesn't, what made a difference to the lives of students and the future of these institutions.

CHAPTER 1

AFFORDING COLLEGE 1980S STYLE: PROBLEMS, CHALLENGES, ISSUES, AND A RESPONSE

In order to fully appreciate the environment in which these educationally purposeful work programs were created and developed, a brief look back to the 1980s is in order.

In the latter part of the '70s, toward the end of the baby boom that had dominated higher education enrollments and produced a period of unprecedented growth, there developed a good deal of support and attention focused on the American middle class, both students and families. The early '70s had been a response to the turbulent '60s. Urban riots symbolized society's unrest. Programs focused and targeted at the poorest in our society (i.e., Affirmative Action programs, the Pell grant—a.k.a. Basic Educational Opportunity Grant—etc.), all part of the Great Society's "war on poverty" were created. By 1975, however, the mood of the country had clearly swung back to the middle. Republicans and conservative Democrats were proposing tuition tax credits for middle-class families. The Carter Administration response was the Middle Income Student Assistance Act (MISAA). MISAA was signed into law in 1978 and the country saw an explosion of borrowing because the "needs test" cap was removed. In 1980, the Reagan Administration set its sights squarely on the reduction of federal support for student aid. Thus began a seemingly relentless series of proposed reductions and even termination of longstanding federal student aid programs budget year after budget year. Social Security benefits, for example, were totally phased out. Efforts to reduce or at least limit the Guaranteed Student Loan program were in part successful. The "needs test," for example, was re-instituted in order for students to be eligible for government subsidized educational loans.

In addition to these funding uncertainties at the federal level, there was a lot more happening in higher education in the 1980s. The quality of our educational system itself was an issue. The report card of our nation's schools indicated that we were behind and losing ground to the rest of the industrialized countries in the world. In addition, 1979 was the high water mark for the size of the cohort of students graduating from high school. The '80s saw a 25 percent decline in the college-going population. Of course, at the same time, the cost of higher education, which then Secretary of Education Bennett blamed from his bully pulpit on the "greedy colleges", was becoming more and more of an issue and the focus of public attention. The cost of higher education and the reduced availability of funds to financially support students already significantly limited choice, skewering enrollments toward low-cost, public university systems away from high-cost, private colleges and universities. Cost was also beginning to limit access because many public institutions were unable to meet increased demand and thus began to limit enrollments. Enrollment management in the 1980s at public institutions took on the meaning of capping or downsizing the number of undergraduates attending. Some of the specific trends of that decade were:

A) In constant dollars tuition in the 1980s increased 52 percent at private colleges and universities, 31 percent at public institutions.
B) The total available student aid nationally in 1989-90 was $30.2 billion. After adjusting for inflation, available aid was 16 percent higher in 1989-90 than it was in 1980-81. However, federal aid in fact declined by 1 percent in constant dollars during this period.[1]
C) Over the course of the 1980s, the federal share of available aid decreased from 83 percent to 76 percent. Nationwide institutional aid grew from 12 percent to 18 percent with state aid hovering between 5 percent and 6 percent.[2] Loans as a percent

[1]Donald A. Wesby and Nancy Carlson, "Trends in Student Aid: 1963-1983." The Washington office of The College Board (December 1983) and Lawrence E. Gladieux, Laura G. Knapp and Roberta Merchant, "Trends in Student Aid: 1983 to 1993." The College Board (1993).

[2]Ibid.

of total federal aid grew from 48 percent in 1980-81 to 63 percent in 1989-90, replacing grant as the most common form of student financial assistance.

D) During the 1980s, the average family income (adjusted for inflation) increased by only 2.2 percent. It actually declined for all but the top two income deciles.

E) The average debt for graduates of high-cost, selective, private institutions was typically $3000 in 1980 and $11,800 in 1990.

F) Significant declines in undergraduate enrollment were being recorded at private colleges and universities. Only 20 percent of undergraduates enrolled nationally attended private universities (down from 50 percent 30 years earlier). More than 50 percent of graduate students nationally enrolled at private institutions. At most private Ph.D. institutions undergraduate tuition subsidized graduate students, thus the future availability of the most highly trained worker, the research scientist or engineer, was in even greater jeopardy in this period of declining undergraduate enrollments.

G) According to *Money* magazine, in 1991, next to catastrophic illness and environmental problems, families rated the cost of education as the most significant obstacle in the fulfillment of their dreams. This poll was taken in September of 1991 when the Persian Gulf crisis dominated the news. Concern about the cost of education was tied for third with the Persian Gulf War.

To summarize, providing prospective students with access and choice to our nation's colleges and universities became more difficult in the 1980s than it had been in the prior 30 years. The viability of many private institutions was called into question, while the public sector was less available to respond to the increasing demand. This was occurring simultaneously with the greying of the professorate and the dramatic increase in business's and industry's need for a highly trained and educated work force. The federal deficit grew so large as to render hopes of federal, financial intervention meaningless; state governments buckled under the new federalism whose mandated but unfunded transfers placed significant responsibility on state governments for social programs (e.g.,the cost of welfare, Medicaid, and corrections, not to mention elementary/secondary education) leaving less for higher education. Given this dismal but unfortunately realistic picture, the only certainty was that business as usual would not get the job done.[3]

During the same period of time, the number of students concerned about making money and getting a job as the reason for attending college rose dramatically. In the early 1970s just half (49.9 percent in 1971) of the new students who participated in the American Freshman National Norms Cooperative Institutional Research program conducted by the American Council on Education and the University of California, indicated that the reason for attending college had to do with getting a job and making more money. That percentage steadily grew so that by the late 1980s it was greater than 70 percent (72.6 percent in 1988). These data, while coupled with the rising degree aspirations and the increased freshman interest in business careers, pointed to a growing student focus on financial security and job opportunities. Students in the late 80s more than ever before were preoccupied with developing their careers rather than using the college years as a time for learning and personal development.[4]

Clearly those "customer requirements" did not go unnoticed by colleges and universities which found themselves in an increasingly competitive market for the best students. The idea of relevance, that is the pragmatic, practical approach to education, became a common and recurrent theme at almost every level of education (elementary, secondary, and post-secondary). Colleges and universities in particular were asked to defend their curricula, especially the liberal arts. There was a widespread perception that the Academy had become aloof, that faculty were "unconnected" with the real world, that the "Ivory Tower" had lost its way.

It was in response to all these concerns, from affordability to utility, that programs like The Cornell Tradition, the Dana funded SAEQ, Duke Futures, and Reach for Rochester were created. They came at a time when pressures on higher education were mounting, as expressed by concerns from access to relevance, from productivity to practicality. In almost every instance as these work-related programs matured, they found it necessary to integrate the work experience into the curriculum, to attach rather than detach, to be consumed

---

[3]Excerpted from James Scannell, "Crisis in the Work Force, Crisis on Campus," *The Admissions Strategist,* The College Board (spring 1993).

[4]"American College Freshman: National Norms," American Council on Education (December 1988).

rather than stand alone, to bring integrity and wholeness to the undergraduate experience through practice as well as theory.

So it was in this environment, with these challenges, that the career-related, educationally purposeful work opportunities emerged. They covered the spectrum from on-campus to off-campus, sponsored by faculty as well as alumni, and in academically related and career-related settings during the academic year and summer. What were the results? What can institutions expect to realize from these experiential education programs? What difference can it make to the undergraduate experience? What value-added and distinctive aspects of an undergraduate education can be realized? What difference will it make?

We have learned a good deal from the thousands of alumni who experienced and benefitted from these programs, some of which was anticipated and hoped for, some of which was totally new and unintended. All of which are of benefit. As a preview to the rest of the story, the following highlights attempt to capture some of the observations, conclusions, and outcomes resulting from these initiatives.

To begin, it was obvious from the receipt of the first response to the written survey that the enthusiasm and zeal of these alumni, and their commitment to their programs and to their alma mater, are extraordinary. Testimony to that can be found in the tremendous response rate (greater than 50 percent) enjoyed in this research, well above what the industry standard would have predicted or what the industry standard would have said was good and significant. Response from this group was double what the statisticians say one should anticipate.

Staying with the characteristics, outcomes, and results the alumni reported, the following was observed:

- Females are more likely than males to participate in these types of programs, that is "self help" programs. In addition, females are more likely to respond to surveys. Thus as you will see in the appendix, there is a compounding effect of females in the response group compared to their representation on these campuses and in these programs.
- Two thirds of the students who participated in these programs were involved as upperclass students only. They had not participated as freshmen and sophomores.
- Students who had a positive experience with a faculty member were most likely and most anxious to share their enthusiasm about the program with others and state unequivocally that it made

a difference. From other responses throughout the research it was strikingly clear that the most dominant, common thread for having a successful experience was a function of whom the person worked with, not what they did. It was the human personal interaction that made the difference and made these programs special. Thus, the faculty apprenticeship, the faculty assistantship, the internship with a faculty member held the most opportunity for a positive outcome of all the various experiences studied. It can hardly be overstated how one is struck by analysis after analysis and testimonial after testimonial of the influence that people had on the interns, the fellows, the assistants with whom they worked and for whom they worked. It was, as stated earlier, not so much what they did, when they did it, where they did it, as with whom they did it. That doesn't mean that in the future these programs will be as successful as they can and need to be by just engaging students in routine work. Clearly the work should be significant and contribute to one's growth and development, both academic and personal. However, the following conclusion is unambiguous. An exciting, career-related, educationally purposeful responsibility without a good mentor has far less value and significance than a more routine and even more mundane set of responsibilities overseen and mentored by someone who cares.

- There was a far more intense, emotional, personal response to an off-campus experience if it was for a not-for-profit, public service, community-based program than if it was in the for-profit sector. There are some logical, obvious reasons for this. These experiences attract the idealistic; students could see that they were adding value to those with whom they worked and the projects on which they worked, for example. We can only speculate that this could well be the first time that these young adults were put in a position of having the opportunity to give rather than to take. That experience itself added a significance and meaning to their work that was not likely possible in the for-profit sector.
- On-campus programs are much more likely to be seen as related to the major than off campus. Again, for logical reasons the academic nature of the setting, as well as many of the assignments, makes an easy connection between what is being done at work and what is being studied in the classroom.
- There were very few experiences (11 percent) that produced a

change of focus in one's major. Why so few given that one could speculate that a placement such as this could just as easily redirect someone into an entire new field and discipline that they had never known or thought of or heard of before? A likely answer harkens back to the previous profile of the group that participated: two-thirds had the experience later in their academic career, that is in the last two years. In all likelihood, majors were well under way by that time.

The factors that contributed to change when it did take place were due to students indicating that they had had the opportunity to develop problem-solving and communications skills, that their setting was in a not-for-profit environment, and that they were placed in an experience that was related to their original intended major.

- Many more students told us that they affirmed their major (37 percent). Of this group, two-thirds said that the experience was related to their major. Also, the experiences cemented major or choice of major. This is more true for on-campus than off-campus placements, where faculty-driven opportunities were more likely the rule than the exception. Again, if you assume that two out of every three students is engaging in these experiences, having already picked a major, it is likely to be much more comforting and satisfying having a real world experience affirm your choice than to send you exploring in the eleventh hour of your collegiate career.

All of this of course begs the question that if one of the goals of these opportunities is to help students sort through the vast array of academic concentrations available to them, it may well be very important to engage students in these opportunities earlier rather than later in their undergraduate careers, clearly before their junior year in college.

- One quarter of the students said that their experiences influenced their decision to attend graduate school. This was not an overwhelming or surprising response. Again it is a somewhat obvious outcome in that most students said that the decision was very much influenced by their working with a faculty member. No surprises here.
- Just less than half (44 percent) of the alumni said that the experience helped them gain professional contacts. Those who cited this as a benefit of participating also cited an increase in their

own self confidence as a result. There will be more discussion about self confidence in Chapter 4. This particular outcome became one of the umbrella or magnet results, that is to say it was one of the developments that when it occurred a lot of other good things happened. There is another strong variable related to the outcome of gaining professional contacts. This result occurred most frequently when the experience included working closely with an alumnus or alumna of the institution.

- Almost half (46 percent) of the alumni said that participating made them feel closer to their alma mater. This particular variable is very much Cornell-dominated because Cornell alumni made up one-third of all responses. However, the testimonial of alumni from all schools, personal interviews with Cornell alumni, and anecdotal data throughout the research clearly indicate that this outcome and sense of community are not something that just happens automatically by exposing students to these work opportunities. The feeling of closeness to alma mater and having a sense of community as an undergraduate are outcomes that have to be developed through a set of conscious and deliberate programmatic initiatives and support with this as a specific goal in mind. There was no proof that just putting students in a good career-related, educationally purposeful work experience will result in their considering themselves so fortunate that they will be forever indebted or connected to alma mater.

  In addition to Cornell there were a number of other programs, Furman and Wesleyan being but two, that worked hard at building a community of learners, workers, and scholars. But it doesn't happen without a good deal of effort. An interesting side note is that although coming into the research we would have predicted that working with an alumnus or alumna would have had a high correlation to this sense of closeness to alma mater, that too was not the case.
- The research produced a clear and unambiguous fact, namely these programs do not in large part improve the academic performance of participants. There was no indication that any particular experience had in any way contributed to enhanced academic performance.
- Graduates of these programs were two to three times (that is 30 percent versus 10-15 percent) more likely to state that they were now involved in their alumni clubs in their local area. This is a

very significant outcome, one that has long-term implications for these institutions or any institution interested in developing a stronger core of alumni. In searching for experiences that would predict this outcome or other results that relate with this benefit, it was of some surprise that closeness to alma mater doesn't match up. On the other hand, a confirming result was that experiences that allowed for the development of a relationship with an alumnus or alumna did predict involvement in alumni club chapters after graduation.

- Response to the question of what skills were developed yielded the following:
  - 40 percent said that an important development was becoming more adaptable and flexible.
  - Nearly 50 percent said that they became more self confident and recognized this as an important character trait.
  - Greater than 25 percent said that learning to work with a team was an important experience for them.
  - More than 45 percent said that developing communication skills was a significant benefit.
  - More than 35 percent cited the development of problem-solving skills.
  - More than 45 percent said that it was really important to become independent and that they did so as a result of this experience.
  - About 35 percent said that they gained technical skills.

  Interestingly, there was no difference in the experience of women at single gender versus co-ed institutions in the development of these skills.

Moving from the general findings of characteristics, outcomes, and results to analysis and conclusions emanating from the nature of the programs, the following are likely to be of value to future program officers and institutional development specialists.

- The dual goals of 1) having these types of programs make an impact on accepted applicants' decisions to enroll at a particular institution and 2) creating the opportunity to reduce indebtedness are very important objectives. However, the experience of these programs demonstrates that they are very difficult outcomes to produce.

As will be seen in the research, Cornell clearly has been successful on both fronts. Only Canisius and Bucknell joined Cornell in reporting a significant impact on the decision to enroll. It was obvious from the research that the promise of future experiential, career-related work opportunities in the upperclass years does not ring true enough or is not "real" enough to entering students to change their matriculation behavior. Successful programs that increased the yield of attendance from accepted applicants had to be specifically geared to what the student could expect in his/her freshman year. This is an important facet of these programs for other institutions to consider if in fact influencing the enrollment decision is an important institutional outcome to achieve. Many institutions in the Dana group, for example, had set this as a program objective but reported no measurable change in enrollment behavior.

Likewise, regarding debt reduction, while many programs also cited reducing indebtedness as an important goal, few were able to execute their activities in such a way that it made a difference. Typically those that did were either programs that touched a small number of students (e.g., Bates) where vast amounts of money were spent on a limited number of participants, or programs that set their sights very specifically on debt reduction (e.g., Furman, Ithaca, Wesleyan, and DePauw) which joined with Cornell in reducing debt for a sizeable number of participants in the program. Forty-six percent of the alumni overall said that participation in these programs had contributed to reducing the level of debt they would have graduated with if they were not involved. It turns out that early participation in the program (freshman and sophomore year) is key to achieving the goal of debt reduction.

- Alumni who participated in experiential career-related work opportunities on more than one occasion during their undergraduate career were 50 percent more likely to feel close to their alma mater than if they participated only once; three times as likely to say the opportunity to participate influenced their decision to enroll at that institution; and twice as likely to say that participation reduced indebtedness. While these are strong and important outcomes, it's interesting that none of the more developmental results related to the number of times one participated.
- It would appear as though there is no particular magic to the

programs offering opportunities in the academic year or the summer or both. Also, no important difference was observed whether it was on campus or off campus. None of these characteristics seem to have a bearing on producing or not producing particular results. Time of participation on the other hand does relate to the development of self confidence and debt reduction. As mentioned earlier, however, all of this paled in significance to whether the experience allowed for the creation of a personal relationship with an adult supervisor, mentor, sponsor, etc.

- There does seem to be something magic about working up to 15 hours a week. The majority of expriences fell between 11 and 15 hours a week. Although not perfectly correlated, the relationship between additional hours worked beyond 15 hours, and satisfaction with the experience, did emerge. The relationship was, as expected, negative; the more hours worked beyond 15, the greater the likelihood that the experience was viewed as unfavorable.
- Many of the participating institutions enroll very affluent student bodies. On more than one occasion, it was noted that these educationally purposeful, career-related work opportunities that provided healthy stipends and/or healthy hourly wages, balanced the playing field for financial aid students. This was not originally an intention in the establishment of The Cornell Tradition; however, as some of the Dana programs emerged it was clear that they were concerned about the relatively small fraction of their student bodies who could not participate, for example, in study-abroad programs, summer internships, etc.—programs which had high educational value but for whom there was no compensation and in fact, even worse, required time and energy which took away from the possibility of paid employment. Needy students in these new programs were able to take advantage of opportunities that otherwise were only available to more affluent populations. Many of the testimonials and anecdotal comments expressed a very real sense of appreciation on the part of needy students at many, many institutions. They saw these programs as having made their educational experience whole.
- Very few institutions were able to sustain the program's development and continuation solely on outside financial support. Almost without exception, institutions (both colleges and universities) had to at least match outside contributions and continue over a period of time to support programs with institutional

funds if they were to extend beyond the period of the initial grant. Of course competition for the philanthropic dollar has only become more intense over the last five to ten years, thus it is likely that institutions looking to develop or expand these type of programs in the future should anticipate as much as half coming from institutional resources.

- Although obvious, it was very clear from the research that only programs having an institutional commitment were ultimately successful. The most successful programs had a champion or champions, that is, the most successful programs had a home, were a part of the institution, were a source of pride, and were owned by the faculty and/or staff. Virtually all of the successful programs became identified with an office or a person, and became a part of the campus culture.

  The analyses demonstrated that the successful administration of the program as well as the requirement to punctuate the individual work experience with a final report did play an important role in the benefit the alumni said they gained from their placement. In addition, of course, the execution of the program positively impacted such things as continuation, fundraising, etc. The extent to which institutions documented and monitored their outcomes seems to have a strong relationship to both successful programs as well as continuation and integration of these programs into the campus culture.

  Finally, in search of the answer to whether these programs make a difference, the data support in a variety of ways (personal, academic, career) that these experiences do make a difference. In fact, the difference isn't only career-making but even life-making. One of the more startling personal outcomes was that of the 2700 respondents, three indicated that they found their mate for life and married as a result of participating in these opportunities.

The attempt in this brief review has been to whet one's appetite to pursue in detail the rest of the text. The goal of this effort is to tell a story that will be useful, not just by accounting for what happened to a handful of institutions who along the way provided some opportunities to undergraduates, but ideally by turning this text into a workbook and a resource. If all goes really well, even a dash of inspiration and vision may emerge.

CHAPTER 2

WORK AND EXPERIENTIAL EDUCATION: ITS ROOTS, VARIATIONS, AND BENEFITS

The roots of experiential learning go back at least to John Dewey's philosophy of progressive education. Dewey promoted the education of the whole person and in the process emphasized the importance of work and manual labor as being a dignified and honorable way in which to obtain the goal. Dewey's notion was simple enough. Guaranteeing universal education as every American's birthright was the problem. How does one system educate the scholar, the searcher, the activist, the artist, the athlete, and the dreamer? How does the American system deliver a learning environment that speaks to all these different types of learners? Part of the answer rests in learning by doing.

Royce S. (Tim) Pitkin, the founder and first President of Goddard College, believed that everybody at college should work at manual labor to make the institution run well. There was no immunity due to wealth. He said, "I never thought that colleges should be havens for the impractical theorists. If we are to survive, teachers and students, they must put knowledge to the test."[1] The college years can be seen as a period of apprenticeship for citizenship. Dewey also noted that building on the interest of students, relating what is done in school to what is done outside and attaining a constant interaction with the wider community, brings value and purpose to what goes on in the classroom. Work programs were one of the ways for students to gain practical experience for living in a democratic society. Goddard College,

[1] Angie Benson and Frank Adams, "To Know for Real," Royce S. Pitkin and Goddard College, page 60.

under Pitkin's leadership, went so far during the years of World War II to add a summer "food for freedom" work camp. Beginning in the summer of 1942, students went to work on farms and in offices and factories to do their share of the war work. They alternated two months of work and study each year. In the 1940s and '50s, working one's way through college was an honorable tradition in America. To work as an integral part of a college education was, however, a new twist.

But Dewey and Pitkin were not alone. William James saw off-campus work and experiential learning as connecting a link between the concept of first and second hand knowledge. Dating all the way back to 1906, the University of Cincinnati organized the earliest Work Study plan. Arthur Morgan followed by setting up a plan of work-related study at Antioch in 1921. In the 1930s, at West Georgia College, Fred Wales, H.H. Giles, and Ed Yeoman used education and resources to help people in surrounding communities. Morris Mitchell pioneered off-campus trips around the country, and even the world, at the Putney Graduate School of Teacher Education and later with undergraduates at Friends' World College. In 1943, Bennington created a non-resident term for working and reading. As will be noted later, Berea, among others, had early off-campus work programs to provide students with experiences across academic disciplines. The history is rich. This chapter will provide a very brief overview of six different variations on this theme.

Experiential Education

A good deal of the motivation and energy for experiential education comes from the challenges as well as the attacks on the aloofness and distance of the liberal arts from the practical world. In the 1970s and 1980s many institutions responded to these charges by creating experiential education programs. Simply stated, experiential education provides the opportunity and the environment for students to experience first-hand, outside the classroom, activities and functions which relate directly to the application of knowledge.

In order for experiential education to pass the traditional academic muster of the faculty, substantiation of its intellectual benefit is a must. One of the ongoing criticisms of experiential education has been the paucity of any scientifically documented outcomes of the benefits of these opportunities as enhancements to student learning. Research as early as 1946 said that experiential education contributed to the aims of liberal education because students could observe society in action; distinguish between good and bad ethics; evaluate the effects of planned social changes or the lack of planning; observe tolerance and intolerance; recognize the difference between creative and passive people; become more sensitive to individuals and groups, and finally understand the histories and theories of human progress.[2] In 1961 Wilson and Lyons reported outcomes of experiential education, linked theory and practice, brought meaning to studies, increased motivation by connecting work and study, and finally developed a sense of responsibility and greater dependence on one's own judgment and maturity. In 1979 Hirsch and Barzak reported students in a field experience found they were active, not passive learners. They took initiative and responsibility, and were more autonomous and independent.

While career development outcomes and personal development outcomes had been documented, cognitive outcomes and learning outcomes had not. Mosser argued that experiential education could have an effect on skills development in a liberal arts education, namely the clarifying of values and integrity; communication skills; critical thinking skills; preparation for work and learning how to learn; cultural sophistication and cross cultural understanding; empathy; tolerance and respect for others; loyalty and intimacy; and finally a sense of self and a social historic context.

Even though the origins of experiential education can be traced to the pioneering work of Jean Piaget, there remains a good deal of skepticism about the value of experiential learning, not the least of which is the concern about vocationalism. Nowhere is the attack on vocationalism more severe than with Benjamin Barber in his recent work *An Aristocracy of Everyone: the Politics of Education and the Future of America*. Barber quite vociferously challenges the vocationalism of the curriculum. He notes, "the vocationalist wishes to see the university prostate itself before modernity's new gods. Service to the market, training its professions, research in the name of its products are all the hallmarks of the new full-service university, which wants nothing so much as to be counted as a peer among the nation's great corporations."

---

[2]John W. Mosser, "Field Experience as a Method of Financing Student Learning and Cognitive Development in the Liberal Arts," University of Michigan (fall 1989).

Barber continues . . . "Education as vocationalism in service to society becomes a matter of socialization rather than scrutiny, of spelling out consequences rather than probing premises, of answering society's questions rather than questioning society's answers. Where once the student was taught that the unexamined life was not worth living, he is now taught that the profitably lived life is not worth examining."[3]

So all has not been rosy for the concept of experiential education. Given the dangers outlined, the risks as defined are often greater than the potential rewards. A balance must be struck and monitored to ensure that Barber's fears are not realized. It is the premise of this study that work and learning not only can mutually coexist, but also can be productive both intellectually and developmentally. Working while learning does not mean that one has to suffer, that one is dominant. Rather, they need to be orchestrated as complementary activities, because one cannot learn completely without working, and of course one won't have worked as productively and effectively as possible if not having learned.

Cooperative Education

Since the 1960s, cooperative education has been a popular and common model for the delivery of professional educational experiences, particularly in the disciplines of engineering and business. It also has served to enhance the productivity of the work force. Institutions like Northeastern University in Boston, Drexel University in Philadelphia, General Motors Institute in Detroit, and many others, have made cooperative education their institutional signature. In its simplest form, cooperative education extends the length of one's undergraduate tenure, from four to five years, by interspersing semesters at work between periods of the traditional academic program. Other variations may not increase the total amount of time spent earning a baccalaureate but imply a 12- rather than 9-month calendar, or require one semester off campus for the cooperative educational experience. As is the case with each variation, the quality of the placement drives the value of the experience rather than the architectural design of the program.

In his article, "Competence, Autonomy, and Purpose: The Contri-

[3]Barber, Benjamin, *An Aristocracy of Everyone, the Politics of Education and the Future of America*, New York: Ballantine Books, 1992.

bution of Cooperative Education," William Weston focused on ways in which cooperative education can contribute to the three areas of personal development originally identified by Chickering. Weston wanted to see whether cooperative education programs helped students establish career objectives and contributed to their own personal development. He proceeded by segmenting the development of confidence into intellectual confidence, physical and manual confidence, and interpersonal confidence. Autonomy included emotional and instrumental autonomy. The development of purpose, according to Weston, was to facilitate a life that satisfies the need for direction and meaning. Weston reported cooperative education as having a positive effect on the personal development areas, which has implications for employers, because they are then more likely to find a satisfied employee.

How does cooperative education differ in content and purpose from field placement? In 1991, Joyce Fletcher in the *Journal of Cooperative Education* compared and contrasted research results on the effect of cooperative education and field experience on a student's personal development, career development, and academic achievement. She found that differences in program goals, structure, students, and placement settings engaged different processes but achieved similar results. Student development—personal, career, and academic—has always been premiered as the biggest benefit realized from a field experience. Field experience holds to the principle that whatever one is doing and however one chooses to do it, the holistic approach to learning produces a significant number of positive benefits. Cooperative education on the other hand has a greater influence on career development. Fletcher found that cooperative educational experiences required greater career commitment, contributed to far more informed career decisions, and produced greater perceived recognition of students' abilities and limitations. Thus, opportunities were more informed, were more realistic, and students tended to have a more realistic view of themselves and what was expected of them. Cooperative education is personal mastery while field experience is more service learning and empowerment. Work experience, Fletcher noted, tends to increase the relevance of classroom work and service learning, even if it isn't career related. Work experience helps academic achievement by enhancing self esteem. All this is to say that the research recognizes different approaches, but with similar outcomes.

Cooperative education to be sure is not for everyone. It is likely to benefit those most focused, most assured, who have a particular career already in mind. It lends itself to being more of a "perfect experiment" than field experience or, as will be discussed subsequently, internships and Work Study.

Internships

By far the most popular of the experiential learning models, internships have a long and noteworthy history in American higher education. Internships have often carried with them the opportunity both to explore the Academy on the one hand, working closely with faculty members on research projects, and to connect to the real world on the other, taking on responsibility in public agencies, not-for-profit organizations, etc. Internships became so popular in the 1970s, the Academy began to search for standards on which to objectively and empirically validate the value, benefits, and attendant requirements of a sound academic internship.[4] Educational benefits of internships have long been noted to encourage more understanding, interests, and participation in government and other civic-minded activities; to observe operating processes; to undertake extensive research; and to gain knowledge of relationships between theory and practice. Successful internships are most often based in and administered by academic departments, strongly supported centrally with requirements linked to the academic enterprise.

The value of internships has always been questioned in the Academy. The critical eye of the faculty has frequently turned jaundiced over them as a legitimate academic exercise.

In 1988, Susan Taylor in her article "Effects of College Internships on Individual Participants" tested three hypotheses. She wanted to know whether the greater the autonomy in the internship, 1) the greater the crystallization of vocational self concept and work values, 2) the less reality shock there would be once entering the real world, and 3) the better the employment opportunities that would emerge. While there was strong support for better employment opportunities emerging, there was only partial support for the greater crystallization of the vocational self concept and work concept and little support that the experience helped with the reality shock of entering the world of work. Overall, it would appear that interns have a distinct advantage

---

[4]1976 National Society for Internships and Experiential Education, Raleigh, North Carolina.

over peers in the labor market on matters such as networking, positive evaluations from job recruiters, offers of higher salaried positions, and greater satisfaction with the extrinsic rewards of the job.

Even though the hard scientific data and analyses always seem to be incomplete, muddled, or contaminated, there nonetheless exists a large collection of outcomes in support of internships as a practical and effective way to begin the transition from theory to application, from classroom to the work place.

Working In College: Work Study And Financial Aid

A good deal has been written, researched, and speculated regarding the value, benefit, effects, and deterrents of working while going to college. The literature is both crowded and confused because there are almost as many reasons why people work on their way to a baccalaureate degree as there are people who work. Thus, a good deal of what is known has to be placed in context or, as Anne-Marie McCartin said in her 1988 article, "Students Who Work: Are They Paying Too High a Price?," the undergraduate work experiences are varied depending on what place on the continuum from work to learning a student is stationed. What does the student perceive as the primary affiliation or commitment? Is the student working in order to make money so that s/he can pay for educational costs or working to learn? Regardless, McCartin says that in her research she has found no "loss of studenthood."

But on that continuum, from working to earn money to working to learn, there are a number of potential benefits and stumbling blocks as discussed by Stampen, Erhan and Thomas. Work may mitigate the transition to full-time work after graduation by serving as an important bridge between college and the work world, even going so far as determining basic employability. It turns out that very few undergraduates actually work to pay for college expenses. Most report working for spending money. Fundamentally, students can't work their way through college today and in fact they haven't been able to for the past 20 or 30 years. In spite of the smaller and smaller part of the term bill being paid by working, the number of students working has increased from 43 percent in 1969 to 51 percent in 1979 to 63 percent in 1991.[5]

---

[5]Carnegie Report in *Change Magazine*, 1968, "The Price of College: Shaping Students' Choices."

Working one's way through school has been and continues to be seen as part of the American tradition reflective of upward mobility. Work has also been seen as having a positive influence on student/faculty integration and satisfaction with one's academic career. Other positives include gaining a sense of independence, having an opportunity to interact more with new people, developing expertise and experience in time management, gaining practical experience that will have applied value in the future. Nowhere in the literature has it been well documented that working has a positive effect on grade point average. As was highlighted earlier, this is consistent with the findings of this research in the over 2700 respondents to the alumni questionnaire analyzed in detail in Chapter 4. The negative side of working one's way through school includes such things as replacing an expected contribution from parents—that is, fulfilling the parents' responsibility for paying for educational costs; the menial nature of work; taking on too much responsibility while in school and therefore distracting one from focusing and concentrating on the business at hand, namely studying; inconvenience in scheduling; and of course sacrificing time from studying or other additional learning opportunities such as socializing or reflecting.[6]

Finally, the literature reports that less than 20 percent of students who work part time while going to school indicate that their work is related to their major or to their career. Those who say it is related tend to be more satisfied with their work experience. These students also tend to be exclusively upperclass students, juniors and seniors.[7] It is in contrast to this outcome in particular that the Cornell, Duke, Rochester, and Dana programs stand apart. While not all assignments were directly related to major or career, they were all intended to be educationally purposeful and substantial learning experiences. It is clear this is not the case with most student work. The differences in benefits as one would imagine are significant.

---

[6]1988 Jacob Stampen, Economics of Education Review, "The Impact of Student Earnings and Offsetting Unmet Need."

—1987 Susan Erhan, Michael Nettles, and others, "Student Financial Aid and Educational Outcomes: Is there a Difference Between Grants and Loans?"

—1990 Debbie Thomas, *The Journal of Student Employment,* "The College Student's Perception of the Effects of Working While Attending School."

[7]1992 Stephen Kane, Charles Healy, Jim Henson, "College Students and

Community Service

While the ideas of working and volunteering as part of one's civic responsibility are not at all new, there has been a tremendous resurgence of late in the context of national service and the formalizing of community service opportunities. In 1990, 26 percent of undergraduates said they were currently performing community service. These students reported working slightly more than five hours a week in such activities. Students who reported that they volunteered at least once during their undergraduate career were more likely to be *women*—36 percent of all female students vs. 27 percent of all male students volunteer (a consistent finding with our research on career-related, educationally purposeful work opportunities). They were also more likely to be *older*—41 percent of students aged 30 and older volunteered compared to less than 30 percent for students under 30 in age. Somewhat surprisingly students who worked were more likely to volunteer than those who did not.[8]

The following brief review captures some of the activity and developments in the areas of service and learning, a concept made popular in the 1980s connecting formally civic responsibility and the acquisition of knowledge.

In 1985, Frank Newman called for higher education to be restored to its original purpose, that is, to prepare graduates for a life of involvement and committed citizenship.[9] Newman proposed that students need to become more actively involved in their own learning. He also raised the issue of linking student aid to community service, the national service concept which has recently been implemented by the Clinton Administration. There has been a good deal of controversy around national service and the linking of student aid to some form of commitment to society. Newman cited the G.I. Bill as being the most effective form of student aid ever developed in our society, a concept based on aid for service. Proponents of community service would argue on the one hand that aid should be earned, that it is not an entitlement. Retractors, however, say that under the present set of

---

Part-Time Jobs: Job Congruency, Satisfaction and Quality," in *Journal of Employment Counseling,* Sept. 1992, Vol. 29.

[8]Eileen M. O'Brien, "Outside the Classroom: Students as Employees, Volunteers and Interns," *ACE Research Briefs,* vol. 4, No. 1, 1993.

[9]Carnegie Foundation for the Advancement of Teaching report, "Higher Education and the American Resurgence."

circumstances only the rich can afford to participate in volunteer activities because the poor have to work. Critics of community service go on to argue that the concept destroys the ideal and the spirit of volunteerism, that in turn it makes needy students second class citizens because the only aid they can get is aid that is tied to fulfilling a commitment.

Regardless, there were a number of community service programs that started in the 1980s, so many that a central clearing house—Campus Compact—was started in the late 1980s by Newman in collaboration with other college and university presidents. The purposes cited for community service include: to engage students in civic responsibility; to connect that civic responsibility with classroom work; to help reduce indebtedness of students while attending college; to have students do more meaningful rather than menial work; and to provide institutions with the response to society's needs, to mention but a few. Interestingly, the programs cited in the literature that have debt reduction as a goal have found how difficult that objective is to achieve.[10] This of course is a similar outcome to the experiences of institutions in this study, as was mentioned in the previous chapter and will be documented in Chapter 4.

Barber provides the best summation for the purpose of community service and community learning in *An Aristocracy of Everyone*. His logic is crisp on the cause and effect: "Learning communities, like all free communities, function only when their members conceive of themselves as empowered to participate fully in the common activities that define the community—in this case, learning and the pursuit of knowledge in the name of common living. Learning entails communication; communication is a function of community. The equation is simple enough: no community, no communication; no communication, no learning; no learning, no education; no education, no citizens; no citizens, no freedom; no freedom—then no culture, no democracy, no schools, no civilization. Cultures rooted in freedom do not come in fragments and pieces. You get it all or you get nothing."

Barber goes on to say, "Historically then, education in America casts training for citizenship as one of its primary purposes. American col-

[10]Christina Kisler, Community Service Program, Westmount College, FIPSE, Washington, D.C., 1989.

—Marvin A. Kaiser, Kansas State Community Service Program, FIPSE, Washington, D.C., 1989.

leges and universities were first founded . . . around the idea of service; service to church . . . service to the local community . . . service to the emerging nations. . . . Service should be understood as a dimension of citizenship, education, and civic responsibility in which individuals learn the meaning of social independence and become empowered through acquiring the democratic arts, then the requirement of service conforms to curricular requirements and the disciplines." He concludes by stating "education is the exercise of authority (legitimate coercion) in the name of freedom: the empower-ment and liberation of the pupil."[11] The ultimate goal, Barber concludes, is not to serve others but to learn to be free which entails being responsible to others. Education-based community service programs empower students even as they learn. They bring the lessons of service into the classroom even as they bring the lessons of the classroom out into the community.

Work Colleges

Work, learning, and service are the cornerstones of five private liberal arts colleges which today remain true to their original mission. Founded to allow poor students from rural areas access to a college education, these early student pioneers simply traded their labor for education. Some stayed to work another year after graduation to finish "paying off" their educational benefits. The five work colleges are:

	Founded	
Blackburn College	1837	Collinsville, Illinois
Berea College	1855	Berea, Kentucky
Warren Wilson College	1894	Swannonoa, North Carolina
College of the Ozarks	1906	Point Lookout, Missouri
Alice Lloyd College	1923	Pippas Passes, Kentucky

Each college is still sustained by its founding commitments:

- to the service of students regardless of their financial means, particularly students from a designated geographical area, usually rural;
- to the value of personal contribution to the campus;

[11]Barber, Benjamin, *An Aristrocracy of Everyone, The Politics of Education and the Future of America,* New York: Ballantine Books, 1992.

- to the egalitarian spirit that is created when all share in work that benefits each member of the campus;
- to the belief that educated people have an obligation to lead and serve;
- to the view that each college is an experiment in fulfilling a radical vision of society born out of an urge to change society, to seek betterment and improvement, and to work toward the perfection of humankind.[12]

Unlike most colleges today, some traditional work colleges actually restrict admissions to students whose families *cannot* contribute above a certain level. Many are first generation college students. These programs truly serve the neediest of the needy. (*See Appendix I* for thumbnail descriptions of each of the five traditional work colleges.)[12]

Two basic types of work programs exist: (1) Those that are an economic tool—work which is on- or off-campus to help pay tuition, and (2) Those with a philosophical commitment to work as part of the educational process itself. When work is intrinsic to the educational process, it becomes part of an institution's reason for existence and commands as much importance in formalities as the academic curriculum. Because the concept of work carries such deep philosophical roots at work colleges, students who do not meet their work requirements are counseled toward the desired attitudes of work and service. While the "buck up or ship out" philosophy lies opposite the work college philosophy, students who fail to meet work requirements after counseling and possible probation could be asked to leave. Some of these schools consider work mandatory, others voluntary. Much of the support and criticism for both parallel the debate noted earlier in national service. There is to be sure no right approach. Whether work is mandatory or voluntary it needs to be monitored, adjusted, retooled to be sure that the objectives and purpose are being met.

Looking more closely at those elements common to work colleges, one finds a deep and abiding commitment to promote a sense of community through work. To create that sense of community and egalitarian spirit found in shared responsibility, one must view work more as a function of self-expression and less as meeting another's demands. Essentially when students develop the habit of viewing

---

[12]"Labor, Learning, and Service in Five American Colleges," A Report to the Ford Foundation (June 1989).

themselves as service contributors to their school, environment, and world, a sense of community evolves. This spirit of community takes root and grows from the seed of good communication. Accurate communication establishes student expectations, provides students a statement of credit on tuition bills, and successfully tells its story through good public relations.[13] Expectations must be clear to students as to how the program works and the philosophy behind it. Most important, students must feel that work is important, that it's not just busy work, and that the school cannot function without them.[14]

This brief review of the American collegiate world of work and learning is meant to serve as a foundation and backdrop to the program descriptions and outcomes that follow. For, although certainly the experiences at Cornell, Duke, the Dana schools, and Rochester share aspects of these work/learn variations, the objectives and designs of the programs at these 23 schools were much more deliberate, defined, and detailed, the outcomes expected more comprehensive, the measurement of the value and benefits more clear.

---

[13]"Report on Student Work Programs Leadership Conference." October 4-5, 1982 (See Appendices II and III, Tables 3-2 and 3-3.)

[14]William R. Ramsay, *America's Work Colleges: Linking Work, Learning and Service*. Unpublished manuscript April 1992.

Chapter 3

Institutional Program Descriptions

The Cornell Tradition, Dana Student Aid for Educational Quality, Duke Futures, Reach for Rochester

In order to provide a programmatic backdrop for the research results, each of the 23 institutions that participated in this project (Cornell University, 20 institutions that received Student Aid for Educational Quality [SAEQ] grants from the Charles A. Dana Foundation between 1985 and 1992, Duke, and the University of Rochester) are profiled. These programmatic descriptions are the composite of original proposals to financial sponsors, yearly or interim progress reports, and an institutional survey which was completed by 20 of the 23 participating institutions. (Dickinson, Parsons/New School for Social Research, and Wheaton provided the final reports that they wrote for the Charles A. Dana Foundation rather than complete the institutional survey designed for this research.) The institutions will be reviewed chronologically by their origination beginning in 1982 with the Cornell Tradition. For ease in finding specific program descriptions, an alphabetical and chronological list of institutions is also provided with their corresponding page numbers. (See Table 3-1a and Table 3-1b.)

A word of caution to the reader. These vignettes are not meant to be all-inclusive. Rather, these descriptions are meant to capture the essence of these creative, educationally purposeful, career-related, paid employment opportunities so that the results of the research, what worked and what didn't and what made a difference to the individuals and institutions can be linked back to each program. Each institution, although somewhat idiosyncratically, recorded, tracked, measured, and evaluated their experiences.

To assist the reader in understanding program distinctives further

Table 3-1a. Alphabetical List

Institution	Page	Date Started
Barnard	84	1988
Bates	43	1985
Birmingham-Southern	66	1987
Bryn Mawr	45	1985
Bucknell	47	1985
Canisius	68	1987
Cornell	38	1982
Davidson	70	1987
DePauw	72	1987
Dickinson	49	1985
Duke	51	1985
Franklin & Marshall	74	1987
Furman	54	1985
Ithaca	60	1986
Lafayette	87	1988
Oberlin	56	1985
Parson/New School	58	1985
University of Rochester	62	1986
Smith	76	1987
Union	78	1987
Vassar	80	1987
Wesleyan	64	1986
Wheaton	82	1987

Table 3-1b. Chronological List

than is afforded in these brief vignettes, a series of tables define the institution by size, type, student/faculty ratio, etc, (Table 3-2); list specific goals of each program (Tables 3-3a and 3-3b); and show at a glance those programs that were offered only on-campus, or were exclusively with faculty, or in the world outside the Academy as well as whether alumni were involved (Table 3-4). Table 3-5 categorizes institutional outcomes as they were defined in their institutional surveys for this research. Lastly, Table 3-6 notes changes made to the original program and Table 3-7 indicates current sources of funding. It is hoped that these tables will amplify the individual vignettes. Please note that the data used to describe the institution in the opening paragraph for each vignette represents figures for 1992.

A review of Bucknell, Canisius, and Cornell programs would be particularly important for institutions wishing to improve competitiveness in undergraduate admissions. A review of Cornell and Ithaca would be instructive if increased alumni giving is important. Powerful outcomes involving students working with faculty abound in descriptions of Furman and Bates, and reduction of student debt was found most significantly at Canisius, Cornell, and Bates. For budget-relieving results and creative use of College Work Study funds, read descriptions for Vassar, Union, and Oberlin. Look also for subtle correlations to where a program found its home. A major imprint was left by the Dana program at Furman, Bates, Ithaca, and Wheaton; the driving forces varied by institution, but similar outcomes were experienced. While this is not meant to be an exhaustive list of what the reader can expect to find in this chapter, it nevertheless provides a map for exploring.

Many paths have been forged by these institutions within the broad guidelines laid out by the sponsors. The purpose of this chapter is to give the reader a sufficient amount of information so that the variations on themes by program and institution, when linked with outcomes, can help others in planning and implementing similar programs.

Table 3-2: Institutional Profiles

	Undergraduate population	Student/Faculty ratio	% Female	Endowment	1992 tuition	Average debt of partici-pants upon graduation	Setting	Liberal Arts (LA)	Pre-Professional (PP)	Known For
Under 3000										
Barnard	2200	12:01	100	56,183,000	14,346	N/A	Metro	LA		
Bates	1537	11:01	53	95,339,000	C	N/A	Sm Town	LA		
Birmingham-Southern	1763	15:01	51	69,439,000	10,200	8,511	Metro	LA		Known as best liberal arts institution in state of Alabama
Bryn Mawr	1278	9:01	100	208,547,000	13,500	10,238	Sm Town	LA		
Davidson	1555	12:01	45	102,441,483	13,150	9,798	Sm Town	LA		
DePauw	2058	12:01	55	140,406,357	10,550	8,520	Sm Town	LA		
Dickinson	2047	10:01	51	78,887,000	13,400	12,609	Sm Town	LA		
Franklin & Marshall	1900	11:01	45	152,999,421	C	17,100	City	LA	PP	Government, premed, business, services
Furman	2480	11:01	53	101,194,000	11,584	5,942	City	LA		Undergraduate research
Lafayette	2045	10.5:01	45	229,047,000	16,725	NA	Sm Town	LA	PP	
Oberlin	2800	12:01	51	240,250,000	17,600	NA	Sm Town	LA		
Smith	2935	10:01	100	435,565,000	13,270	10,877	Sm Town	LA		
Union	2013	9:01	44	117,551,537	13,513	11,626	City	LA	PP	
Vassar	2218	10.5:01	58	268,630,000	16,250	9,166	Sm Town	LA		
Wesleyan	2800	11:01	50	319,449,000	14,610	11,919	Sm Town	LA		
Wheaton	1319	10:01	65	58,410,000	8,350	12,984	Sm Town	LA		
Between 3000-5000										
Bucknell	3300	13:01	46	136,082,000	13,825	7,729	Sm Town	LA	PP	
Canisius	3617	18:01	45	35,000,000	7,552	3,657	Metro	LA	PP	
Parsons/New School	3468	10:01	49	37,641,000	13,066	11,000	Metro		PP	Parson's premier programs in art and design
Rochester	4900	13:01	48	656,178,000	13,425	11,336	Metro	LA	PP	Engineering
5000-10,000										
Duke	6130	24:01	46	669,075,000	13,153	13,928	City	LA		7th in nation
Ithaca	6126	12:01	53	91,060,000	12,870	9,038	Sm Town		PP	Music
Over 10,000										
Cornell	12585	N/A	45	1,078,340,643	16,214a	10,384	Sm Town	LA	PP	

[a]Endowed; N/A = Not available; C = Report comprehensive tuition, not tuition alone

Sources: *The Chronicle of Higher Education Almanac*, Issue August 25, 1994. Individual Institutions Represented

Table 3-3a: Institutional Goals: Chronological Listing

	To remain competitive	To maintain enrollment	To increase enrollment	To increase alumni giving	To instill moral obligation to perpetuate program	To increase alumni/ student interaction	To increase student/ faculty interaction	To increase faculty productivity	To reduce debt through work and recognition	To increase summer employment	To free students' career options	To provide career-related experiences	To make students more competitive/employable	To increase academic year opportunities	To provide educationally meaningful paid internships	To increase opportunities in non-profits	To increase research opportunities	To make students more competitive in the marketplace
Cornell	X				X				X	X	X	X	X					
Bates								X	X						X			
Bryn Mawr									X	X				X	X			
Bucknell	X		X				X		X						X			
Dickinson									X						X			
Duke						X				X		X				X		
Furman						X	X		X	X		X		X		X		
Oberlin							X	X							X			
Parsons/New School												X	X					X
Ithaca				X					X			X			X			
University of Rochester									X			X		X	X			
Wesleyan									X						X			
Birmingham-Southern									X			X			X			
Canisius		X						X	X						X			
Davidson									X			X						
DePauw									X			X				X		
Franklin & Marshall									X						X			
Smith												X		X	X			
Union									X			X					X	
Vassar							X								X			
Wheaton									X			X			X			
Barnard									X						X			
Lafayette	X								X						X	X		
TOTAL	3	1	1	1	1	2	4	3	18	4	1	12	2	4	16	4	1	1
1985	1		1			2	3		2	3		2		2	3	2		
1986				1					3			2			2			
1987		1					1	1	5			5		1	4	1	1	
1988									1						2	1		

Table 3-3b: Institutional Goals: Alphabetical Listing

	To remain competitive	To maintain enrollment	To increase enrollment	To increase alumni giving	To instill moral obligation to perpetuate program	To increase alumni/ student interaction	To increase student/ faculty interaction	To increase faculty productivity	To reduce debt through work and recognition	To increase summer employment	To free students' career options	To provide career-related experiences	To make students more competitive/employable	To increase academic year opportunities	To provide educationally meaningful paid internships	To increase opportunities in non-profits	To increase research opportunities	To make students more competitive in the marketplace
Barnard									X						X			
Bates								X	X						X	X		
Birmingham-Southern									X			X			X			
Bryn Mawr										X				X	X			
Bucknell	X		X				X		X						X			
Canisius		X						X	X						X			
Cornell	X				X				X	X	X	X	X					
Davidson									X			X						
DePauw									X			X				X		
Dickinson									X			X			X			
Duke						X				X		X				X		
Franklin & Marshall															X			
Furman						X	X		X	X		X		X		X		
Ithaca				X					X			X			X			
Lafayette	X								X						X	X		
Oberlin							X	X							X			
Parsons/New School																		X
University of Rochester									X			X						
Smith												X		X				
Wesleyan				X					X		X				X			
Wheaton									X						X			
Union									X			X					X	
Vassar							X								X			
TOTAL	2	1	1	2	1	2	4	3	16	4	2	11	1	3	14	5	1	1

Table 3-4: Program Profile

	Time of Year			Freshman	Sophomore	Junior	Senior	Alumni							
	Academic Only	Summer Only	Both					On Campus	Off Campus	Both	Yes	No	TA	RA	Faculty/ Admin.
Barnard	X				X	X	X			X	X			X	X
Bates			X					X				X			
Birmingham-Southern			X	X	X	X	X			X	X		X	X	X
Bryn Mawr			X	X	X	X	X			X		X	X	X	X
Bucknell			X	X	X	X	X	X				X	X	X	X
Canisius	X			X	X	X	X	X				X		X	X
Davidson			X		X	X	X			X		X		X	X
DePauw			X		X	X	X			X		X		X	
Dickinson			X			X	X	X			X	X			X
Franklin & Marshall			X	X	X	X	X			X	X			X	
Furman			X		X	X	X			X	X		X	X	
Ithaca			X		X	X	X			X	X		X	X	X
Lafayette			X	X	X	X	X			X				X	X
Oberlin				X	X	X	X						X	X	X
Parsons/New School			X			X	X			X	X				
Smith			X	X	X	X				X	X		X	X	
Wesleyan		X			X	X			X		X				
Wheaton			X	X	X	X	X			X	X				X
Union			X		X	X	X	X				X	X	X	
Vassar			X	X	X	X	X			X		X		X	X
Duke		X		X	X	X			X		X			X	
Cornell				X	X	X	X						X	X	X
Rochester			X	X	X	X	X			X	X			X	

Table 3-5: Institutional Outcomes*

	Reduce student debt of top students	More competitive graduates	Increased student involvement on campus	Improved student/faculty interaction	Improved faculty productivity	Improved quality of faculty	Increased giving opportunity for donors	Increased reputation for institution	Model for other programs	Relieved campus budget for college work study	Showed institution other means for student support other than term time (summer)	Provided extra administrative help to faculty	Administrative staff more satisfied	Increased enrollment	Improved quality of student body/ intern pool	Positive impact on prospective students	Source of high-quality students for specialty campus jobs	Increased university and business interaction - good relation	Increased alumni student interaction	Recognized value of student participation in faculty research	Increased support for student participation in summer research	Engendered student appreciation	Source of financial aid
Barnard																	X						
Bates																							
Birmingham Southern	X			X								X											
Bryn Mawr			X	X				X					X										
Bucknell				X										X	X								
Canisius	X				X																		
Davidson				X					X														X
DePauw					X												X	X					
Dickinson					X								X				X			X			
Franklin & Marshall								X															
Furman		X		X	X	X		X															
Ithaca				X			X											X	X				X
Lafayette				X																X			
Oberlin				X	X					X													
Parsons/New School																		X	X				
Smith					X					X						X							
Wesleyan				X														X	X			X	
Wheaton					X				X									X	X				
Union										X											X		
Vassar										X	X										X		
Totals	2	1	1	9	7	1	1	3	2	4	1	1	2	1	1	1	3	5	4	2	2	1	2

* *Dana supported schools only.*

Table 3-6: Changes to Original Program

	Not based on financial report	Increased award amount	Reduced minimum number of service hours required	Exclusively faculty/student collaborative work	No trainee program	Reduced number of participants	New model - initiative	Independent projects eliminated	Merged with existing internship program	Increased hourly rate	No longer give board stipends	No longer exists	Abandoned entrepreneurial experience	College does not subsidize student stipend	Now one of other research opportunities with faculty	Fewer academic year opportunities	No changes	Dropped academic year program - retained summer	Reduced size of summer program
Barnard						X													
Bates	X					X												X	X
Birmingham-Southern	X			X		X	X												
Bryn Mawr																			
Bucknell		X								X									
Canisius																			
Davidson							X												X
DePauw							X					X							
Dickinson																			
Franklin & Marshall											X			X					
Furman					X														
Ithaca						X											X		
Lafayette	X										X				X				
Oberlin	X					X													
Parsons/New School						X													
Smith									X										
Wesleyan						X													
Wheaton																			
Union																		X	
Vassar								X	X									X	
Duke												X							
Cornell		X	X																X
Rochester													X						
Totals	4	2	1	1	1	7	3	1	2	1	2	2	1	1	1	0	1	3	3

Table 3-7: Current Source of Funding

	Internal	Foundations	Gifts	Endowment	Outside source grants	Sponsor pays cost of each student	Alumni (annual giving)	Corporations	Federal
Barnard	X	X					X		
Bates		X	X	X1					
Birmingham-Southern	X			X					
Bryn Mawr				X					
Bucknell				X2					
Canisius	X	X			X				
Davidson		X	X						
DePauw		X							
Dickinson	X								
Franklin & Marshall						X			
Furman	X								
Ithaca	X			X3					
Lafayette	X			X	X				
Oberlin		X							
Parsons/New School	X	X						X	
Smith				X4					
Wesleyan	X	X	X	X			X	X	
Wheaton	X	X	X				X	X	
Union	X	X						X	
Vassar				X	X				
Rochester	X				X				X5
Duke - extinct									
Cornell			X	X			X		

1 = Permanent for program
2 = Endowed scholarship funds
3 = Program endowed
4 = Dana Challenge Grant
5 = Very limited

The Cornell Tradition, 1982

Cornell is unique among Ivy League schools in that it is not entirely a private university. State tuition funds three undergraduate schools: College of Agriculture and Life Sciences, College of Human Ecology, and School of Industrial and Labor Relations. Because of its public and privately funded colleges within the university, Cornell boasts one of the most diverse student populations, registering about 25 percent minority students. Located in Ithaca, New York, Cornell dominates this small town. Cornell's endowment is, of course, the highest among the institutions in this research, at over $1 billion. Founders of the Cornell Tradition embraced the founding philosophy of the institution:

> "I have always been in favour of combining labour and study."
>
> Ezra Cornell, 1846

Ezra Cornell founded his University with a commitment to provide students with opportunities for work as well as study. The Cornell Tradition is the seminal career-related, educationally purposeful, paid student employment program. In fact, it is the model that sparked Dana Foundation interest in SAEQ as well as the programs at Duke and Rochester. Not only because The Cornell Tradition is the oldest of the programs reviewed in this research; not only because more than one third of all the alumni in this study are graduates of Cornell; but because The Tradition, from its beginning, was by far the most thorough in documenting, tracking, measuring, evaluating, and analyzing experiences and outcomes, it is virtually impossible to capture all the facets, components, and nuances without dedicating an entire work to this now 14-year effort. Thus, again by way of caution, this review will only highlight The Cornell Tradition in order to provide a backdrop and reference for what follows.

The Cornell Tradition began in 1982 with an anonymous $7 million gift. The original program proposed six key features. The program sought to bolster recruitment efforts for undergraduate students by offering the opportunity for career-related summer employment to continuing students regardless of need and to reduce the indebtedness of graduating seniors in order that they might more easily pursue careers in which they were most interested. This would be particularly true for those who wished to pursue graduate study or service-related careers. In addition, the program sought to improve the employability of Cornell graduates by making career-related job opportunities available during their undergraduate program. Lastly, the program intended to instill in Tradition Fellows the moral obligation to perpetuate The Cornell Tradition through future service and financial assistance to the University and to The Tradition Program.

In the beginning there were four specific Tradition programs: Freshman/Transfer Fellows, Academic Year Fellows, Summer Job Network, and Summer Fellows.

The Freshman/Transfer Fellows Program was a direct recruitment initiative matched with the first goal of keeping Cornell competitive for the best students. During the admissions process, school and college admissions committees[1] nominated Tradition Fellows based on their demonstrated commitment to the work ethic and community service while in high school or for transfer students during their collegiate years. Essentially, freshmen and transfer applicants who demonstrated high academic talent, a commitment to community service, and high financial need were eligible for an award of up to $2000 a year originally (eventually growing to $2500) that reduced by that amount the indebtedness a student would incur obtaining a Cornell degree. Tradition Fellows were, and are, strongly encouraged to participate in the program throughout their undergraduate years so they will have substantially less indebtedness upon graduation than non-program participants. The research supports the finding that multiple year participation significantly reduces student debt. The Cornell Fellows model is built on the premise that students will participate all four years. Students, however, reapplied each year. Besides awarding fellowships, the Tradition enhanced student employment and internship opportunities as well as community service experiences. The program also helped Fellows develop closer ties to the University through contacts with donors who have endowed Tradition fellowships. (Because of the strong four-year commitment to selected students, alumni of Tradition programs are expected to provide subsequent generations with job opportunities and financial support and be actively involved through various volunteer activities.)

[1]Cornell has seven undergraduate schools and colleges, each with its own separate Committee on Admissions: College of Agriculture and Life Sciences, College of Architecture/Art, and Planning, College of Arts and Sciences, College of Engineering, School of Hotel Administration, College of Human Ecology, and School of Industrial and Labor Relations.

Freshman yield rates for accepted candidates chosen as Cornell Tradition Fellows ran anywhere from 10 percent to 20 percent higher than overall yield rates for the freshman class. For a view of Dana funded programs modeled on the Cornell Fellows goal to enhance recruitment efforts, see the program description for Bucknell and Canisius. Both of these Dana-funded schools focused on recruiting freshmen and provided work opportunities throughout the undergraduate experience to help reduce student debt.

To remain eligible for the Academic Year Fellows program in subsequent years, students originally had to demonstrate 360 hours of paid work in the prior academic year, maintain a minimum 2.3GPA, and participate actively on campus. It was found that the work expectation in the early years was too steep and was subsequently reduced to 330 hours in the second year of the program and eventually dropped to 300 by 1985. The application and selection processes were extremely rigorous and comprehensive. The first participants in the Academic Year Fellowship program numbered 170 Fellows in 1983-84. That number grew to 197 a year later. Eventually the goal of 600 Fellows from both the Freshman and Transfer as well as Academic Year programs was met. Each participant received an award, initially set at $2000 a year which reduced by that amount their indebtedness.

To assist and reward students who choose community service or other low-paying but educationally purposeful, career-related work opportunities during the summer, the Summer Fellowship was designed to supplement their required summer earnings. Often, if students chose these lower paying, but community service-type jobs, they couldn't meet their $1200 summer earnings requirement. Additionally, those who took summer jobs away from home in order to have a meaningful career-related experience, used their summer earnings for living expenses and typically finished the summer with little or no savings. Although the Tradition anticipated a need for 160 summer fellows, the demand for this program never reached that level. Summer Fellow numbers hovered near 100 annually.

The fourth component to the Cornell Tradition was the Summer Job Network, which was the only non need-based portion of the program. The idea of the Network was to provide linkages for undergraduate students to exciting career-related, educationally purposeful opportunities in their own hometown and around the country. In order to stimulate the marketplace, the Tradition provided a subsidy of 40 percent to the for-profit sector and 70 percent to the not-for-profit sector in order to create opportunities. Those subsidies were subsequently reduced slightly in 1985-86. Applications for the Summer Job Network were staggering almost from the very beginning with 1340 applications in 1983 and almost 2000 in 1984. All applications were individually reviewed and an assessment made of the student's skill level with the various fields requested. Students were then notified of the kinds of jobs to which they might be referred. In 1984, 2600 multiple referrals were made for positions with 565 students placed for a placement rate of 30 percent. The Summer Reach program at Rochester is an almost exact duplicate of the Summer Job Network.

Although one out of every four undergraduates at Cornell applied for Tradition programs in 1987, introducing the program to the rest of campus (faculty, staff, etc.) in order to create an understanding and appreciation of the impact and the potential of The Cornell Tradition remained an elusive goal. There was concern about dual efforts of the Student Employment Office which ran Cornell Tradition, and the Career Services Offices of the various colleges. A concern was also expressed about the difficulty in having freshman and transfer Fellows meet the academic year requirements of work hours. Finally, and most importantly, there was a desire to have The Tradition fully integrated into the mainstream of the University so that it would become a part of the typical Cornell undergraduate experience.

In 1987, after five years of experience with The Cornell Tradition, a comprehensive review was conducted which led to significant changes. As a result of the review, the following substantive changes were made. First the work requirement for Academic Year Fellows was reduced from 300 to 200 hours for freshmen and from 300 to 250 for sophomores and juniors. The review revealed that students struggled to meet the work hour requirements and instead of helping students it became a major stress factor. There was greater recognition given for community service and campus involvement. It was deemed important for the continuity of participation for incoming freshmen or other new students to understand that the award could continue for four years or for the duration of their undergraduate career. Four-year awards were given to freshmen who had to maintain certain requirements and document participation through a simplified application process. In addition, students could participate for the first time as upperclassmen. The Summer Job Network was spun off to the Career Service and Placement Office and the subsidy program terminated. Eligibility for job placements limited to Cornell Tradition Fellows.

To more fully integrate The Tradition into Cornell's mainstream, in 1988-89 a faculty adviser was appointed, dedicating 20 percent of his time to The Tradition. Finally, the reporting structure changed from the Student Employment Office, which reported to the Director of Financial Aid, to a stand-alone office reporting to the Dean of Admissions and Financial Aid. (Reporting and operational structures are noted for several of the Dana Programs, especially when they contributed to significant outcomes such as at Ithaca.)

In 1988-89 a student advisory council began with a goal of building a stronger sense of identity with the Tradition Program and sense of community among participants. Although its early activity was spotty, by 1990-91 the student advisory council was firmly established with subcommittees including faculty outreach, alumni relations and fundraising, public relations, volunteer network, Fellows directory, phonathons and special events, tee shirt design/production committee, and the "Hello Fellow" monthly newsletter. In addition to the student advisory council and the faculty advisor, other activities were initiated such as student employee job fairs, a breakfast for the parents of Tradition freshmen during Parents' Weekend, a homecoming reception for Tradition alumni, a senior barbecue for Tradition participants on Commencement Weekend, a Cornell Tradition display at reunion, named Fellows reception at the Trustees' Council weekend, Cornell Tradition alumni news letter, and college-based receptions for Fellows. These outreach and community building activities worked toward building a sense of spirit, community, and fellowship among The Cornell Tradition participants. These community building activities proved to be a cornerstone to the participants' affinity toward their alma mater.

Finally, with an eye toward career development, an executive-in-residence program was begun in 1991. This program called for a notable chief executive officer to spend three days on campus with students in small group seminars focused on career development and related current professional topics. In addition to the seminars, large group presentations, receptions, and a student advisory council dinner filled the agenda for those three days. This now has become an annual event. Wheaton College features a similar event.

Today The Cornell Tradition annually awards 600 Fellows and has an alumni population of over two thousand. The Fellowship program boasts an 86 percent program retention rate for individuals throughout their undergraduate career. Three full-time, two part-time, and two student workers staff the Tradition programs. This represents an increase of one full-time worker over the original staffing level. The Cornell Tradition has enjoyed extensive national print and electronic media coverage including CBS Evening News and *US News and World Report* to name but two. Fundraising is ongoing with over $22 million having been raised in the first ten years of the program ($225,000 from foundations, $35,000 from corporations, $8.7 million from alumni, $25,000 from parents, and $13.3 million from other individuals). A minimum of community service hours has been established and many affinity group, community building activities, as noted earlier, have been added.

Today the Tradition remains a vibrant and dynamic set of program opportunities for undergraduates. With each graduating class, The Tradition demonstrates how programs of this kind can make a difference to the institution and its service to future generations of students as well to individual participants.

Bates College, 1985

Located in the industrial city of Lewiston, Maine, population 80,000, Bates College is 2½ hours from Boston and offer a traditional liberal arts curriculum. Bates has the distinction of being the first co-ed college in New England. Almost 90 percent of the students hail from outside Maine, many from Massachusetts. Bates is almost the smallest institution in this research at 1537 students, second only to Wheaton College. Their endowment is nearly $100 million. Similar institutions in size of population, that is, under 2000 students, include Franklin & Marshall and Bryn Mawr.

The Dana Faculty Research Apprenticeship program at Bates is one of the most powerful models for students working with faculty. Indeed, the imprint on the institution created a legacy for a permanent, yet smaller, faculty student collaboration. The program was designed to broaden access to a private education for financially needy students who otherwise would not imagine that a private liberal arts education could be available to them. Funded from 1985-1990, it was in the first group of institutions funded by Dana. Bates took seriously the idea that theirs would be a model for others to follow, and in fact report many inquiries about their program.

The program called for the entire financial need of participating students to be met with grants in the third and fourth year of their undergraduate experience. Research apprentices would work with faculty starting in their junior year and continuing until graduation. The purpose of the single focus on research with faculty was designed to enable faculty to increase their productivity while helping students reduce debt through work in an educationally meaningful experience. Over half of all full-time faculty were expected to participate over a five-year period. The Dana program was intended to provide an alternative to traditional merit scholarships. The smallest of the Dana-funded programs, it funded 39 students during the grant period.

The budget called for $500,000 to be spent on student tuition remission; $50,000 for advisers; and $50,000 for room and board in the summer, travel, and printing expenses. The average was set at $17,000 per student for financial aid, summer work stipends, and room and board. Students were not paid directly except for the $500 summer stipend.

Qualifications for the program were perhaps one of the more stringent of the Dana funded programs for both students and faculty. In order to qualify for the program (two years tuition remission and $500 summer stipend), students had to meet financial aid criteria, be involved in campus life, provide three references, and have an interview with the faculty sponsor. Faculty proposals were reviewed by an advisory panel, consisting of one faculty member from each division. The Office of the Dean of Faculty advertised the selected proposals to students. Students who applied were interviewed by the faculty sponsor. Other criteria considered were students' GPAs. Students who completed these projects also wrote a formal report.

Because the program spanned two academic years with a summer experience sandwiched in between, this program provided continuity of assistance to faculty. The full-time summer experience was expected to compensate for the start-up training time of the first year. Bates' proposal stated that this program was "an unusual opportunity for faculty to engage in technical or narrowly focused projects of intense interest to that scholar and to enjoy the companionship of a supporter." Otherwise such research, they felt, can tend to isolate faculty, not only from students, but even from departmental colleagues. From a student point of view, the program provides a substantive experience that introduces one to research and the collegiality of the Academy.

Although alumni were not involved in developing opportunities or supervising students, there was a 17 percent increase in the parents' annual fund the year the Dana program was featured in fundraising messages to parents. The College also reported that in 1992 annual giving at Bates was 100 percent greater than it was in 1984. There was an obvious significant debt reduction of more than $5000 for students who participated in their junior and senior years. The program had no impact on majors because by the time the students participated they had already found a home for their academic concentration.

Bates stated that in the future, faculty and students in the humanities and social sciences would need to be targeted, as their participation in the first five years of the program was very limited. Housed in the Dean of the Faculty's office, the Bates program was supported by one staff member who carried other duties as well.

Since the Dana grant expired, the program continues at one quarter support—that is 6 versus 25 interns with combined funding from the College's current operating budget. The program in its second life after the Dana grant was modified to be a summer only program and not one dependent on financial need.

In recent years there has been a great upsurge in interest among faculty and students at Bates in student/faculty summer research projects, as well as service internships. This growth in interest is in part due to the legacy of the successful Dana program. Bates reports receipt of new grant funds and personal gifts for their program from the Andrew W. Mellon Foundation, Maximilian O. Hoffman Foundation, Howard Hughes Medical Institute, and the Mulford Foundation, some of which are endowments. To underscore the importance of student research at Bates, over 60 students stayed on campus over the summer of 1994 to participate full time in faculty-guided research or in service-learning internships. (Bates student body is 1525 total.)

Bryn Mawr College, 1985

Located on 135 acres in Philadelphia's wealthy main line suburb, Bryn Mawr College houses 1850 undergraduates and graduates in a peaceful and self-contained setting. The nearby train station provides easy public access to Philadelphia. Bryn Mawr was the first of predominantly women's institutions with extensive graduate programs to grant a Ph.D. to a woman. Bryn Mawr is one of the smallest institutions funded in the Dana program. Bryn Mawr's endowment is over $200 million.

Bryn Mawr College's Dana Internship Program served 253 students over a four-year period. The Program was designed to achieve two primary goals: first, to make available to bright, financially needy students jobs that would enrich their education; and second, to provide more employment opportunities at Bryn Mawr for all students during the academic year and summer, thereby encouraging greater numbers of able students to consider Bryn Mawr, while lightening the loan burden of financially needy students.

The Dana Internship Program allowed students the opportunity to develop skills closely related to their academic and career goals. The program offered a variety of skilled jobs including teaching assistantships in tutorial sections, administrative internships, research assistantships, and professional externships. All of the jobs were planned to be intellectually challenging, work opportunities that both complemented students' coursework and allowed them to test various career possibilities. The majority of jobs in the Bryn Mawr Dana Internship Program were designed by the students themselves and were awarded competitively.

The Dana Internships were seen as awards, not jobs. Students were funded with stipends, not salaries. Summer support included a $1200 stipend plus room and board; in the academic year they were granted $1800 stipend. Almost all of the 261 participants were juniors and seniors. Undergraduates who were eligible for financial aid were solicited to apply for the program. All applications were reviewed by a committee of administrators and faculty. Other than financial aid criteria, proposals selected demonstrated a close match between a student's academic program and her career goals.

By 1990, Bryn Mawr had raised almost $400,000. Five internships a year are now supported with endowed funds. The program boasts a 99 percent retention-to-graduation rate on the part of participants versus an 85 percent rate overall. The Dana Program has been used in admission recruiting through a special brochure; however, there has been no evidence of any measurable enrollment results due to the program. The program encourages faculty to work with students as colleagues. Examples of on-campus activities were: intern for a chemistry professor working on inorganic synthesis; intern for a curator of college collections for care, conservation, and cataloging of vintage photographic materials; and an intern for a professor working with materials from an archeological dig.

Examples of off-campus activities were: intern for the Academy of National Science of Philadelphia in Education Program; teaching

ESOL with the Lutheran Settlement House Women's Program; intern to the Director of Homeless Outreach Project; and a position with the Volunteer Lawyers Program of Berkeley Community University Center.

Bryn Mawr is one of a small but significant number of Dana institutions that struggled with and felt burdened by the administrative monitoring, tracking, evaluation, and reporting of the Student Aid for Educational Quality Program. The program was housed in the Financial Aid Office and staffing was provided by two part-time workers and one student worker.

This reporting concern is an issue which sponsors and program directors continue to struggle with as programs emerge at institutions with different cultures, structures, sizes, and visions. To be sure, there is a tension and perhaps a healthy balance between the program goals and objectives and the measurement and monitoring of activities and outcomes. The balance needs to be struck based on resources, institutional philosophy, and sponsor requirements. Bryn Mawr was not alone in this concern.

Bucknell University, 1985

Bucknell University, located in the small town of Lewisburg, Pennsylvania, falls in the mid-size category of institutions funded by Dana at 3300 undergraduates. The student/faculty ratio is 13:1 and the endowment is $136 million. Bucknell is defined primarily as a preprofessional school, since it offer majors in business and engineering.

Bucknell and Canisius, like Cornell, used the Dana Program to recruit top quality undergraduates to their institutions. Bucknell students were guaranteed meaningful work opportunities beginning in the freshman year as were those at Canisius.

Used principally as a recruiting tool, the Dana program at Bucknell assigned interns to tasks meant to provide them with experiences to enhance their educational growth. Each participant was mentored by a faculty member or administrator. The program in turn also provided valuable high-quality assistance to faculty and senior administrative staff. Although Bucknell was in the first group of institutions funded by Dana and those proposals represented the most broadly based goals, their goal was perhaps one of the most clearly defined, evaluated, and successful in meeting the stated objectives of the program. Success can also be measured by program continuation, which Bucknell's did with slight modifications.

There were four basic changes to Bucknell's original program. Summer internships were phased out after the first year. Program administrators found that students who were appointed to summer positions tended not to continue as Dana student interns thereafter. No specific reason was given for the drop-off in participation. Since Bates was successful in incorporating a summer experience between the junior and senior year, perhaps it was too early developmentally for some students to remain on campus immediately following their first year.

The second change was also related to the program's phase-in process. During the first year funding, Dana interns were appointed for all four class levels. After the first year of funding, only freshmen were appointed. The third change had to do with changing faculty directors. Faculty directors of the program were granted one course of release time during the spring semester of each year. In addition, the faculty director was provided an annual stipend of $3000. The last program change affected the amount of the students' grants increasing from $900 to $1500 and the hourly rate from $3.70 to $7.50, with a $1500 cap.

The selection of Dana interns begins with the Dean of Admissions who selects a large number of superior and/or talented students who will be admitted to the University. Selection criteria includes a combined SAT score of 1250 (pre-recentering) or greater, placing in the top 5 percent of the graduating class, a record of leadership involvement in voluntary service, and an interest in improving the environment in which they live, and/or minority status. A financial aid officer then screens the candidates to determine who will require financial aid. This narrowed field of potential Dana candidates, usually between 150-200, is then given to the Dana Director to select between 20-50 students who are offered the Dana internship. The Dana Director writes a special letter to each student describing the internship program and inviting them to accept Bucknell's offer of admission. They are invited to an on-campus open house. Not all candidates accept the offer of admission and not all who accept the offer of admissions choose to become a Dana Intern because of the obligations involved.

For those students who accept the offer, each receives a $900 grant that would otherwise be student loan. In addition, Dana Interns are accorded greater than $900 in wages to be earned over the two semesters. Dana interns are very attractive to departments because they only have to pay one-third toward the student wages with the other two-thirds provided from central funds.

Matching students to projects is done by the Dana Director. The Director invites proposals from faculty and administrative offices and selects those opportunities that promise a significant degree of mentoring, direct supervision, and educational growth to the student. The interns' qualifications and interests are reviewed for both incoming freshmen and returning students to find the best match. The demand for interns exceeds the supply, and returning students may continue in the same assignment. All participants are required to make a formal report on their internship.

There was a serious attempt to build community at Bucknell among the Dana Fellows. Typically in the fall there was a formal dinner as well as lunches held every other week with faculty, staff, and Dana participants. In the spring there was a formal dinner and a cultural event.

Bucknell has considered using the Dana program framework to target specific populations beyond those presently served (i.e., the most highly credentialed students). For example, similar internships could be offered to minority students or non-traditional transfer students whom the institution wants to attract to the student body.

Finally, the "Bucknell Approach"—that is, entry as a freshman; internship status for four years; a work experience that is oriented to student growth; selection and supervision by a faculty member; careful placement in the work environment; group activities for interns; etc.—was noted as having worked very well for the institution. The program enriched the education of those students who were interns. It offered high-quality assistance to faculty on University projects and administrative operations. It added to the attractive features of Bucknell that taken together persuade students to enroll.

The Bucknell program continues today at about the same level of participation. An endowment restricted for this purpose is the source of grant support.

Dickinson College, 1985

Dickinson College is located in the small-town setting of Carlisle, Pennsylvania, on 130 acres. The 2047 undergraduate student population enjoys a student/faculty ratio of 10:1. Dickinson is considered a small, private liberal arts college that also offers 3-2 programs in engineering with Case Western Reserve, University of Pennsylvania, and Rensselear Polytechnic Institute. Students may also arrange classes with member institutions of the Central Pennsylvania consortium. Dickinson's endowment is $78.8 million.

Among the first institutions funded under the Dana SAEQ series of grants, Dickinson proposed to expand and strengthen their efforts to provide adequate financial assistance to help students meet their educational costs, and to help students see the relationship between their present educational experience and their life and work upon graduation. To accomplish these goals, the program included an Alumni Parent Program in which Dana interns would be placed with parents and alumni for summer employment; Academic and Administrative Assistants who would work as research and teaching assistants to faculty and assistants to Dickinson College administrative staff.

One of the most distinctive components of the Dickinson program was the Dana Community of Scholars. The Community of Scholars program provided a forum for participants to socialize and exchange useful information. Assistantships produced interesting results that other supervisors and assistants enjoyed learning about. The presentations themselves allowed assistants a chance to refine their communication skills. Participation in the Community of Scholars program was voluntary but more than 50% of the assistants and about a third of the supervisors attended on a regular basis. The Dana experience at Dickinson was an upperclass program—no freshmen allowed—the majority of internships took place during the academic year on campus as opposed to summer internships.

Dickinson reported that the Dana Student Assistantship program was one of the few ongoing programs on the campus that promoted faculty research opportunities and one of just two programs that involved undergraduates in this process in a systematic way. More than 275 academic and administrative assistantships were funded over the program's first five years involving over 225 students. The scope of the work included assistance with scientific research and experimentation, data collection analysis, software development, bibliographical research, statistical compilation and writing, contributions to publications, assistance with curriculum development and revision, research and publication of in-house brochures, policy statements, and news stories. Dickinson reported that its faculty have come to see the Dana program as indispensable to their research. A large majority of student respondents see the program as contributing to or at least influencing their career decisions. The Dickinson evaluation revealed that supervisors agreed that the program had a significant impact on the quality of teaching and learning and that the program improved the quality of their own teaching. Faculty received a stipend for working with students during the summer. The program was open to upperclassmen.

Regarding programmatic changes, Academic-Administrative Assistantships were opened to all students regardless of financial need. Dropping financial aid criteria was the most common change to programs when Dana funds ended.

Dickinson also reported that while the program did not reduce significantly the amount of student debt (most of the participants simply replaced another Work Study position on campus for their assistantship), it did alter the way the campus viewed student work. Because the Dana program provided students with jobs that are educationally meaningful, faculty and administrators in general have elevated the level of responsibility of many of the other Work Study positions on campus. Therefore more students are working in jobs that have more long-term relevance and benefit than before. Since, as mentioned above, the program affects the work portion of a typical financial aid package, Dana funding allowed for the replacement of a different kind of employment on campus rather than supplementing existing aid awarded to individual students. The SAEQ program added to the total number of jobs available on campus, which had a positive impact on the total financial aid program, but did not alter materially the amount of aid that an individual student received. Dickinson now supports the program at the $140,000 level.

Duke University, 1985

Duke University certainly enjoys wide name recognition. Located on two beautiful and spacious campuses in the major city of Durham, North Carolina, Duke is sometimes called the "Harvard of the South." The 6000 or so undergraduates are divided between two divisions of the University: the School of Engineering and Trinity College of Arts and Sciences. Duke enjoys one of the larger endowments in the nation at $669 million. Duke, like Cornell and Rochester, was not a recipient of the Dana SAEQ series of grants. Though internships were most closely related to a student's academic interests, a key aspect of the program was emphasis on alumni involvement with undergraduates in a positive and supportive way.

First named the Duke Futures Scholar Intern Program, Duke's best undergraduate scholars as well as other "hard to recruit" populations such as North Carolina students and minority students were targeted. After two years the program was basically opened up to all Duke undergraduates and was changed simply to *Duke Futures*. The goals of the program included helping students find alumni-sponsored, summer internships that were relevant to their academic study or intended careers. In addition, for students on financial aid, Duke wanted to be able to provide jobs in non-profit organizations without students worrying about their summer savings requirement. Finally, Duke hoped to encourage a more active and productive relationship between the University and its alumni in preparation for a Capital Campaign that was to be launched in the next two years.

The Duke program was noted for placing interns in very creative, exciting summer work experiences, often with an alumni connection. (A small number of on-campus summer research positions were also available in the Medical Center.) One Duke alumnus noted that the Duke Futures experience led to his current job and although his formal internship only lasted a month, he and his mentor are still working together after five years. As a result of his work he has received many industry awards (Mac User Best of 1990; Group Ware 1992 Best of Show; ITCA Best of 1991; *Discover* Magazine Innovation Award, 1992; US Patent, 1993). As a participant in the Duke Futures Program, this alumnus created the first simultaneous conference software for the Macintosh that gives users and particularly students in an educational setting the power of a collaborative writing lab. This rightfully proud alumnus noted the firm with which he is now a senior officer has hired three Duke grads and two Duke interns in the last five years.

Connections between alumni and undergraduates were the heart of the Duke Futures Program. Over 1200 alumni were cultivated by a staff of five full-timers. The majority of these alumni had no prior involvement with the university and those that had, routinely complained that the institution had only contacted them for money. They were delighted to contribute back to their alma mater in what many considered to be a more personally satisfying way. Alumni not only provided internships in their businesses for undergraduates, but also formed teams to develop jobs in major cities.

Students were recruited for the program by word-of-mouth, faculty advising, pre-major advisers, flyers, and ads and articles in the campus newspaper. No minimum GPA was required, but a demonstrated willingness to work hard was essential. A resume and an interview with program staff was required and students had to re-apply each year. Employers in the profit sector paid the entire student's salary, while those in the non-profit sector were eligible for student wage subsidies. The program was very popular with employers (large and small) who viewed it as a cost effective recruiting tool and positive PR for their firms. Only a small portion of the program was faculty directed. Approximately five to ten internships a year took place in the Medical Center in various areas of research. To fulfill the goal of recruiting freshmen to the program, Duke Futures staff also traveled for admissions. Publicity for students selected into the program included a press release to students' hometown newspapers.

Although separate, a Service Learning Project developed as an outgrowth of Duke Futures through a Fund for the Improvement of Postsecondary Education (FIPSE) grant from the Department of Education. That program supported 10 to 15 students per summer who wished to pursue full-time, community service internships. Support was given directly to students in the form of a stipend (after analyzing their budget) rather than as a subsidy to organizations. Students on need-based aid were eligible for additional awards to replace their summer savings expectation.

In 1990-91 the character of the program changed dramatically and Duke Futures became part of the Career Development Center. The referral system was dropped as were the job development teams and most of the procedures, goals, and objectives. The program officially ended in 1991 after serving a total of 800-900 students over five summers. Those involved with the program were cautious to speculate on whether there was any real longstanding benefit. However, from the students' perspective, the first major benefit was the work experience and potential job contacts the program made possible for those who participated. Certainly their earning power was enhanced, a fact clearly demonstrated in this study's research. The second major benefit was the availability of internships in the non-profits made possible through the subsidy/stipend and summer savings replacement components of the program.

Duke participants in this study showed a significantly higher rate of participation in alumni clubs. In fact, no other program except Cornell demonstrated such increased participation in alumni club activity. For institutions interested in increasing alumni participation, involvement in the career development of undergraduates works. Other programs with summer experiences to compare would be Davidson, Wesleyan, Rochester, and Cornell. The Duke students frequently made the case that their experience in this program definitely influenced their choice of major and career.

Furman University, 1985

Located on a 750 acre campus about five miles from downtown Greenville, South Carolina, Furman University is a selective co-ed, independent institution. A liberal arts college with strong pre-professional programs, the University's 2480 undergraduates enjoy an 11:1 student/faculty ratio. Furman emphasizes independent research and internship opportunities, supports a community service program involving nearly 1400 students, and has university-supervised foreign study programs in seven countries. Furman students hail from 42 states and 18 countries. The University's 1992 endowment was $101,194,000.

The Furman Advantage Program is one of the strongest success stories from the Dana SAEQ series of grants and had perhaps the boldest proposal. The transmittal letter from the President of Furman University said that if Dana supported the program for the first four years, then they would take it over. Dana did. Furman has. It is now known as "the Furman Advantage." At the heart of the Furman program was the establishment of on-campus research fellowships for rising seniors in which students began full-time work during the summer before their senior year assisting faculty members with research projects and teaching fellowships. In addition, the project supported off-campus internships and educationally related summer traineeships developed with the help of Furman alumni. All aspects of the program were closely related to participants' academic studies and career goals.

The formidable imprint on the institution directly touched both students and faculty and indirectly contributed to an enriched culture on campus. In fact, faculty testimonials in a 1987 interim report to the Dana Foundation were extraordinary. Over the course of the grant, more than half of the faculty were involved. Of those, over 95 percent claimed that the research assistant had a significant impact on their research, and that the quality of the students was high, and urged the Furman administration to support the program after the Dana grant expired. Having a Dana Intern became a status symbol and each year more faculty applied than could be accepted. Students shared in the prestige and benefitted also. One hundred percent of the students who participated as research assistants were accepted to graduate schools compared to the average Furman rate of 88 percent. The Furman Advantage Program grew to its current status on campus not only by being a solid program, but by including spirit-enhancing activities also. The program sponsored picnics, volleyball games, and recognition dinners for participants. A new tee-shirt is designed each year advertising the program.

The program was coordinated from the Dean's Office and was directed cooperatively by the Director of Financial Aid and a faculty member. This partnership between the faculty chair and Financial Aid Director provided essential program administration and technical expertise. These two representatives, a committee representing each academic area (humanities, social sciences, sciences, and arts) together with the Director of Educational Services, selected faculty and student participants. Even though it may sound cumbersome, Furman says it works. Starting in 1993, an Assistant Dean handled the administrative coordination.

A further examination of the process used in the on-campus research program noted, the faculty member's application form required an analysis of the project, the exact duties of the intern, the publication expected as a result, the qualities needed in the research fellow, a vita, and a one-page letter of recommendation accompanying the student application if the faculty member had a particular student in mind. In almost all cases, the faculty knew the students they wanted. Clearly, the Furman Advantage research fellows and teaching fellows programs are controlled by the faculty who submit to one of the most rigorous selection processes found among the Dana funded institutions.

Students applied directly to the program only for off-campus internships. These required a description of the position from the employing agency and a letter of support from a Furman faculty member. Typically 12 to 14 summer interns were placed and six to eight placements occurred during the academic year. In addition, the Dana summer trainee program consisted of alumni-developed, off-campus opportunities for upperclassmen related to their major. The goal was to have 65 to 75 traineeships per year where the companies employing the students provided the salaries. Furman had a very aggressive institutional goal to provide a career-related summer job to everyone who applied for such a position by the 30th of November of the preceding calendar year.

As noted above, the research fellows were clearly the focus of the program. The purpose of those positions was to increase students' future opportunities, enhance faculty development and morale, and to make the University attractive to more good students. There were 25-35 research fellows each year over the four years of Dana funding. Dana research fellows must have had a 3.0 GPA or better. They received $1600 for their work as well as housing over the summer. Approximately 20 Dana teaching fellows assisted faculty in teaching five hours a week each academic year. Teaching fellows, who needed to have a 3.0 or greater GPA in their major and a demonstrated financial need, received an $800 stipend for their teaching assistantship.

The 1987 report to Dana stated that 15 percent of the students reduced debt an average of $1600. Furman sought greater debt reduction among more students. There was not, as of the time of the interim report, a measurable impact on admissions. Eighty-eight percent of the participants said they would be excited to stay involved with the program. There have been new corporate internships and corporate support, many joint publications by faculty and students, and a $250,000 challenge fund from a foundation. Presently the program is funded directly by the institution's operating budget and through fundraising.

The goals of the program have changed only slightly over the years. The traineeship program was dropped. As the program matured, the summertime off-campus program was dropped so more attention could be focused on on-campus opportunities where there was a more direct relationship to the student's academic program with faculty supervisors. The benefits of what has come to be the "Furman Advantage" include, for the institution, the ability to attract, retain, and revitalize faculty; provides an edge, sense of identity, and specialness for the University; and creates a greater sense of academic community. Furman very much wants to endow the program. Retention through graduation has been 100 percent for those participating compared to a campus-wide 90 percent enrollment retention. It is truly a program that has had a significant impact on the campus community. Furman is a different place because of the Dana SEAQ grant.

Oberlin College, 1985

Oberlin College's 440 acre campus in the small town of Oberlin, Ohio, an hour from Cleveland. This small college of about 2800 undergraduates includes a world-renowned conservatory of music. Over one-fourth of the students are minorities, perhaps one of the more diverse campuses in this study. Oberlin, which is the leading college in the United States for Ph.D. production, was first to grant degrees to women and the first to declare its instruction open to all races.

Oberlin was in the first group of institutions funded in 1985 under Dana's SAEQ series of grants that shared broad-based goals. The program opportunities were mainly on campus.

Oberlin proposed a program of Undergraduate Research and Teaching Assistantships. Oberlin said the program would respond to the needs of students for summer and campus employment with specific relevance to the curriculum, and would strengthen education by bringing students into sustained contact with faculty members who are active in research by providing increased opportunities for participation in research and teaching. Since Oberlin does not have graduate students, the program focused on supporting faculty with student development as a byproduct of the interaction. This program is comparable to the one at Bates College. Both programs included a summer and academic year experience on the same project. (Bates actually included two academic years with a summer experience in the middle.) Perhaps it was the extended contact with faculty at Bates that provided a catalyst for a permanent program and a shift in educational philosophy. By contrast, at Oberlin the Dana Program was used to supplement CWS funds that targeted specific subpopulations of students based on major and/or ethnicity. The program expired at the end of Dana funding.

Oberlin's program was administered from the Provost's Office. Faculty projects were selected by the Research and Development Committee for educational value and overall quality and selected faculty solicited students directly to work with them. An occasional position resulted in students seeking out specific faculty members to work with. Basically, students learned of these Dana internships opportunities directly from faculty, or by word of mouth through other students; no formal publicity was conducted.

To be eligible for the program, students' financial aid packages had to include a campus job award or student loan and had to have completed at least one intermediate or advanced course pertinent to the area of research. Stipends were set at $3250 for research assistantships (full-time summer, part-time academic year) and $1500 for teaching assistantships (part-time academic year). These funds allowed students to meet the College's expectations for summer and academic year earnings as well as allow some reduction in cumulative indebtedness. Oberlin saw the Dana program as being particularly appropriate because of its own institutional distinctiveness versus research universities; Oberlin promotes its undergraduate experience as personalized by close interactions with a world-class faculty.

Oberlin reported that there were 400 program participants overall—302 unduplicated—distributed fairly evenly over four years. One-quarter of the experiences took place during the summer, three-quarters during the term, all on campus. No credit was provided for the experience and no alumni were involved. The typical Oberlin Dana experience on campus included research assistantships in the chemistry department working on transition metals, and research and teaching assistantships in the history department supporting classes on modern American history. Off-campus work with faculty included ecological field work in Colorado with the biology department, social surveying in a nearby county with the sociology department, and field-based research in Latin America with a professor from the biology department.

After the Dana funds expired, 30 summer and 30 academic year internships have been maintained with College funds and outside support from such benefactors as Howard Hughes Medical Institute, BP America, Mellon, and Ford Foundations. The new programs have not been restricted to financial aid students. Oberlin noted the Dana program was seen as a benefit to faculty research as well as a positive experience for the institution in its preparation to apply for other such grants. Some of the opportunities in the successor programs are limited because the funds are restricted, for example to science students in the BP America grant, minority students in the Mellon grant.

Parsons New School for Social Research, 1985

Parsons School of Design is the largest college of art and design in the country with 3468 undergraduates. Located on a two-acre campus in the heart of New York City, it was founded in 1896 and merged with the New School for Social Research in 1970. The institution has always believed in merging the academic and the practical, professional world of work. Frank Alvah Parsons, the founder, is noted as being the first American educator to see the relationship between the visual arts and industry.

Building on its founding philosophy and education practice, Parsons created the Charles A. Dana Foundation Student-Industry Partnership, which called for an expansion of existing pilot projects at the School. In these expanded Dana projects, students training for careers in visual arts supplemented their classroom experience with hands-on internship experience in the industries related to their academic training. The Dana program was also designed to expand the amount of financial aid currently available to students. To this end, approximately 25 percent of the new internships established with the Dana funds were to be reserved for students admitted to Parsons under the auspices of the Higher Educational Opportunity Program (HEOP), a program for educationally and academically disadvantaged students who are residents of New York State. The Dana program thus was not only targeted for financial assistance but also intended to identify a group of very talented but very economically needy students. Participation in this program was expected to increase their employment potential upon graduation.

Although the program was initially to be marketed to freshmen, as the program evolved it was not included in freshman admissions literature because the School didn't want to raise expectations from a large number of unqualified candidates. Even though Parsons did not market the program as aggressively as planned to incoming freshmen, two main components provided great strength: it increased alumni involvement through many off-campus internships and thus created closer ties to industry. By the end of the grant period, 225 placements had been made over a four-year period. The placements occurred primarily in the junior and senior years. They were almost evenly split between summer and academic year and wages were approximately $7.15 per hour with average earnings over $2600. To qualify, students needed to have upperclass status with an adequate portfolio, be in good academic standing, and demonstrate need for financial aid. Program coordination was handled by a half-time equivalent administrator.

Parsons noted that the Dana program provided the following benefits to the institution and to students: Direct industry experience was provided to departments that could not have before participated—that is, the departments outside the previously piloted areas of the crafts and fashion design areas. The new departments with internships included communications design, photography, environmental and interior design, and illustration. In addition, Parsons started an advisory board for each of the seven departments. The program also opened up internships to international students in the Bachelor of Fine Arts program and created the opportunity for foreign internships in the Far East. Given the competitive world of art/design Parsons boasted a high employment rate for graduates of the Dana experience in 1989, greater than 20 percent in companies who sponsored the placement opportunity.

Success extended to funding as well. Parsons attracted the following financial support: $793,000 including two $100,000 grants from foundations; $458,000 from business; and $135,000 institutional support. The project continues today with Parsons covering the full cost. Although there are fewer internships offered (about half the number that occurred during the Dana grant) the Dana program served to expand internship opportunities to new disciplines and strengthened industry ties.

Ithaca College, 1986

Ithaca College is the largest private residential college in New York State with just over 6000 undergraduates. With slightly more women than men, students are enrolled in one of five schools that make up the College: the School of Humanities & Social Sciences, the School of Business, Roy H. Park School of Communications, the School of Health Sciences and Human Performance, and School of Music. There are just 100 full-time graduate students. Ithaca College, along with Cornell University, contributes one-third of the population of Ithaca, whose total population is 70,000.

Funded in the second wave of Dana grants (circa 1986), Ithaca's proposal was perhaps one of the strongest multi-dimensional proposals and the only program, besides Cornell, that focused on and was successful at significantly increasing alumni giving and other fundraising. In addition, their proposal targeted freshman and transfer student recruitment and retention, reduced indebtedness, enhanced the undergraduate experience through academic and career development, and involving alumni and faculty in work experiences. Ithaca's program is one of a few that did not change its program goals at the conclusion of the grant period. More importantly, Ithaca's program, like Furman's, became a signature program for the institution.

With strong institutional commitment to the program's continuation, Ithaca took an aggressive approach to fundraising from the start. They accomplished this, for example, by including the Director of

Development on the oversight committee. The program remained an institutional fundraising priority. Alumni giving nearly doubled over the course of the grant period and is attributed, in part, to the Dana grant. (Fifteen percent of Dana alumni have made contributions in the past five years.) The program now enjoys the benefit of a $152,000 endowment which came from the original fundraising and approximately $75,000 a year support from the College. How did Ithaca raise the money? In general, alumni solicitation occurred directly through a series of ads placed in the quarterly alumni magazine and indirectly through feature articles. Dana program alumni will continue to be solicited in the future, adding to an already growing base of endowed program support. Ithaca was able to attract new corporate and foundation support for their institution as a result of the Dana program however, it was noted that results from corporations were less than anticipated. The internship program continues but on a smaller scale with 25 interns a year rather than the 60 that was the rule during the years of the grant.

Administered from the Provost's Office by an oversight committee consisting of the Directors of Career Services, Development, and Financial Aid, the program was made known to students through extensive outreach including campus newspapers, radio and TV stations, evening information sessions, word of mouth by supervisors and interns, and individual mailings. Applicants to the program were judged on an interview with program staff, a writing sample, a transcript, a completed application, and references. Essentially, students chose either from a placement that already existed, developed their own experiences, or were matched depending on the project. Likewise, faculty underwent a selection process. They were judged on the quality of the proposed internship and their record of supervision. Off-campus employers were recruited through announcements distributed by the College or were directly recruited by the students. The oversight committee made final decisions on matches between student and projects.

The program called for term time academic year stipends that were renewable, as well as summer work internships which could be held on a one-time basis only. All participants would be called Dana Interns. Incoming freshmen and transfers as well as continuing students were eligible but had to reapply each year. Fifty-nine faculty sponsored the academic year interns, which were primarily on-campus, while alumni and parents sponsored the summer interns. The variety of experiences included on-campus placements in academic departments, and off-campus placements at both not-for-profit and profit organizations. At the conclusion of Dana funding, a total of 300 renewable academic year experiences and one-time summer work internships were completed.

A seemingly small, but contributing, factor to a participant's overall satisfaction with the experience had to do with recognition. Although a reception for participants and sponsors was dropped after the first year, each participant continued to receive a letter of commendation upon receipt of his or her internship evaluation. Press releases were also sent to the student's hometown newspaper, high school guidance counselor, and parent(s) and/or guardian. These hometown news press releases about upperclass interns provided an indirect recruitment benefit for the college.

The Dana SAEQ program became a signature program for Ithaca College, with fundraising one of the jewels in the crown. Over the course of the grant, the College raised $62,000 from corporations, $65,000 from par- ents, $103,000 from other sources, and $171,000 from corporations and foundations, for a total of $401,000. A good deal of the funds from individuals came from the Dana Interns Sponsor Program. A program's success depends on more than fundraising, though. Ithaca had a well conceived plan for recruitment, selection, and variety of opportunity.

University of Rochester, 1986

Located in Upstate New York, near Lake Ontario, the University of Rochester is one of the smallest private research universities in the country. With under 5000 undergraduates and a student/faculty ratio of 13:1 students can find an academic home in one of six colleges that comprise the University. This predominantly liberal arts university also offers undergraduate programs in engineering and nursing. The greater Rochester community is home to three major corporations, is the 13th largest exporter in the country, and is the major imaging center for the world. The Reach for Rochester program shares the philosophy and some of the goals of The Cornell Tradition.

Reach for Rochester is a creative, multi-component approach to student financial assistance that maximizes student earnings while providing career-related, educationally purposeful work experiences. Reach emphasizes the concept of self-help and social responsibility mostly through summer employment.

The Student Employment Office was the original sponsor for the Reach program, but in 1991 the Student Employment Office and the Career Service and Placement Center merged to form the Center for Work and Career Development (CWCD). Through ongoing research, the CWCD has found that work-related experience emerges routinely as the single most important factor in securing a first job. In fact, a recent survey of employers by the Center noted that job readiness and experience was mentioned most frequently as the attribute and qualification for hiring into entry level positions. (Forty-two percent of all employers ranked this attribute in the top three.)

The Reach program focuses on the need to expand and enrich student employment opportunities for educational, financial, and career development reasons. Reach began with three programs. First, Reach Experienceships —paid, project-oriented work during a student's sophomore and junior academic years, on and off-campus. Students participate both in upper level administrative projects as well as in research activities with faculty. Off-campus, students participate in community service projects. Students earn $1500 a year and if in a community service placement, an additional $2500 scholarship to reduce their indebtedness is included. The Reach Experienceship has been billed as a $4000 a year part-time job.

The second feature of the program was Summer Reach for which students applied in the fall for placement the following summer. Students prepare a resume and writing sample and are interviewed in the Center for Work and Career Development. Teams of alumni, parents, faculty, staff, and friends conduct job development activities throughout the year. A typical Summer Reach job paid $3000 of which $1000 was subsidized by the Reach program. The partial subsidy is required in order to generate the most exciting, career-related, educationally purposeful work opportunities. Of the 553 placements which occurred between 1989 and 1993, three-quarters of the participants were upperclassmen. A standard applicant pool for the Summer Reach program includes 450 applicants, 1100 referrals, 200 openings, 100 placements—15 at no subsidy, 10 at 25 percent subsidy, and 75 at a 40 percent subsidy. Students typically earn up to $3500 over the summer.

The third component of Reach, Reach Enterprisers, was phased out after three years because of lack of student interest. Seed money from the University was provided to students to develop an approved business concept. Local alumni provided counsel as did faculty from the business school. Students earned a percent of the profits. Projects included special recognition "memory clocks," a floral service, etc.

The desired outcomes of the Reach program are quite similar to those of The Cornell Tradition, The program strives to make available opportunities that positively affect students' career decisions and that also serve as an invaluable complement to a liberal arts education. Participation is known to increase the potential of being hired by an employer who places a premium on work experience. Similar to the self-help philosophy of The Cornell Tradition, the Reach program is an opportunity for students to make financial contributions to their educational costs thereby reducing indebtedness, which may expand career choices. Through its many opportunities in the non-profit sector, the program strives to create good citizenship and make significant contributions to society through community service. It is marketed as providing an opportunity to create your future and build the leaders of tomorrow. The program seeks a greater presence among employers, alumni, and the community at large.

There is a good deal of alumni participation, as noted by five recent reunion classes having restricted gifts to the Reach Program—the Classes of '37, '52, '61, '86, and '88. Despite all the program successes, there has been only minimal effect on the reduction of student debt. However, program participants have experienced significantly better retention through graduation than is typically the case in the student body as a whole.

Wesleyan University, 1986

Wesleyan University, located in Middletown, Connecticut, is home to about 2800 undergraduates and approximately 150 graduate students. Wesleyan, a liberal arts university, is among the most highly selective institutions in the country. The student/faculty ratio is approximately 11:1. Wesleyan's student body is fairly diverse with an even split between males and females and approximately 35 percent students of color. Wesleyan is a fairly affluent institution with an endowment at over $300 million.

Wesleyan established The Dana Summer Scholars Program for students who demonstrated need and who were eligible according to a set formula for grant aid and Federal College Work Study funds. The program was confined to the summer because both the Admissions and Financial Aid Offices, as well as Career Planning, said that was where the program was most needed. The only change to the program since Dana funding ended is a decrease in the number of participants due to limited support. Similar to the Summer Reach Program at Rochester, Wesleyan's program focuses on one-time summer experiences. The program enables students to apply for internships that help them think more clearly about their educational aspirations and how to realize them. The goal as stated originally was to ensure that all undergraduates have reasonably comparable educational experiences, that student indebtedness be reduced and associated constraints on graduate study and career options relieved, and that the program stimulate existing efforts underway in student services, fundraising, and alumni affairs. Indeed, alumni were involved in providing rewarding summer work experiences.

Summer placements were both on and off campus with stipends ranging up to $2500. Summer earnings were applied to the summer self-help expectation, and summer living expenses. Dana scholars' summer earnings resulted in a $500 loan reduction on any need-based Wesleyan loan for the academic year following participation in the Dana program. An average $500 loan reduction was recorded by each of the 122 participants for a total of $61,000 less indebtedness. Students could hold Dana Summer Scholarship placement only once during their undergraduate career.

Housed in the Career Services Office, the program was made known to students through the Career Planning Newsletter, direct mailings, student newspaper articles, and information sessions. Selection criteria included a proposal, writing sample, resume, transcript, and financial aid forms. The strength of the proposal was most critical and was evaluated by a faculty committee. The committee looked for proposals that would enable an intern to acquire new or additional skills, or an enhanced educational experience that would broaden areas of career interest. Off-campus positions were developed by job development teams, but the student was responsible for soliciting a specific employer for his/her proposal. Wesleyan believed that the process of applying for the internships would help students think more carefully about their educational aspirations and how to realize them. Finally, Wesleyan committed to raise money for this Dana program through a $10 million capital campaign, alumni giving, and corporate gifts. Because of the job development recruitment aspect of the off-campus program, 13 part-time staff and 10 student workers supported the effort.

By 1990, three years into Dana funding, 122 students had participated in the Dana program, most of whom were sophomores and juniors. The average earnings were $2674, the range from $1025 to $5450. Virtually all of the positions were off campus (99 percent). Wesleyan's internships could best be classified as follows, using the examples from the thirty 1986 placements: one research assistantship with a Wesleyan professor off campus; 11 internships in public service/not-for-profit organizations; one internship in a corporation; two apprenticeships with artists; seven internships with government agencies; four scientific research or laboratory internships; two publishing internships; two journalism internships; and two internships in small art agencies. There was no course credit given, but a formal report was submitted.

There was a heavy service orientation to many of the experiences. For example, one student was a teaching aide and counselor for the Kola Achievement School at the Pine Ridge Indian Reservation in South Dakota, another an aide to a nurse midwife and courier for the Frontier Nursing Services in rural Appalachia, Kentucky; and finally, another student was a researcher in the areas of plastic recycling and solid waste management for the Northeast/Midwest Institute, a nonprofit policy research organization in Washington, D.C. Wesleyan promoted the program in its admissions campaigns; however, there were no measurable recruitment results provided.

Wesleyan plans to continue the Dana program at about half the level (15 versus 30 interns) noting that the program needs to be supported by endowment, not operating funds. Wesleyan raised over $235,000 between July of 1985 and December of 1986 in support of its match for the Dana program.

As planned, alumni were also targeted for a gift solicitation specifically for the Dana program. As a result of this effort, the Wesleyan Black Alumni Association Council endowed an internship. Significantly, alumni contributions were also noted by Davidson, Cornell, Duke, Ithaca and Rochester alumni for their respective programs.

Birmingham-Southern College, 1987

Birmingham-Southern College, located in Birmingham, the largest city in Alabama, was created by a 1918 merger of Birmingham and Southern Colleges and is affiliated with the Methodist Church. Although it is known primarily for liberal arts, more than one third of the students major in business-related areas. The 1700 or so students enjoy a student/faculty ratio of 15:1. This small southern school has an endowment of $69 million. Birmingham-Southern offers a Master of Arts in Public and Private Management. (New in 1995: a Master of Accounting.)

In the third wave of Dana-funded institutions (circa 1987) Birmingham-Southern proposed a program of financial assistance responding to the need for additional student aid and the desire to provide enriching opportunities for undergraduates. The program was built on an existing experiential learning project of on-campus, academic year work geared to bringing a closer linkage between education, work, and community service, while at the same time bringing a dimension of the work ethic to a liberal arts education for those students who participated. The Dana program opportunities were both on campus and off, in the business community and with civic organizations.

During Dana funding, the program was housed in both the Dean's Office and Career Services. Students were made aware of the program through brochures, mailings, signs, word of mouth, and reputation of the program. Students interested in becoming Dana interns completed an application, a writing sample, and an interview with program staff. Students were selected based on GPA, recommendations, skills, and experience. Once selected, they were matched and placed by the program. There was no selection criteria for faculty sponsors; any faculty member could volunteer, and over 60 faculty members were involved over the grant period. As in many programs, those students completing an internship wrote a formal report, and completion of an internship was recognized by the institution at appropriate awards ceremonies. Students interested in another Dana-funded work experience had to reapply. About a quarter of the participants were freshmen, the remaining upperclassmen. Typical of academic year work, students worked 10 to 15 hours a week at pay above the minimum wage. Average earnings during the academic year were just over $1200, while average earnings in the summer were over $2200.

Placements under the original Dana grant included: research assistants (10 a year, mostly rising juniors and seniors), "Docents," teaching assistants (10 a year, again mostly juniors and seniors), lab assistants and tutors, and administrative interns. Specific placements included a lab tutor position in biology, an administrative intern in business and economics, and a research assistant in statistics. Typical off-campus placements included such assignments as a counselor's assistant in the Path Program (an agency for homeless women), an administrative intern, an assistant law intern responsible for paralegal work, and finally, a marketing intern who assisted with the writing and editing of brochures for the local symphony.

When Dana funding ended, students were no longer selected on the basis of financial need. Assignments were less focused on educational value, but more targeted to finding students with specific skills for jobs on campus such as math lab tutors, computer assistants, and research assistants. Birmingham-Southern, having benefitted from the Dana experience, formed a task force to review undergraduate research. The Task Force proposed the College Fellows Program (CFP), which consisted exclusively of joint faculty/student collaborative work, research, and teaching opportunities. This program matches students and faculty on teaching, research, and curriculum development projects. Students are selected based on merit and, like many other institutions with continuing or evolved programs, financial need was no longer part of the selection criteria. The CFP focuses exclusively on faculty-student collaboration, and serves about half the number of interns as the Dana Program (35 versus 70). The current source of funding is an existing scholarship fund on campus. Thus, similar to the outcome at Bates, the Dana program at Birmingham-Southern became the seed from which a new program that now focuses on educational value was formed.

Canisius College, 1987

Located in Buffalo, New York, Canisius College is an independent Roman Catholic (Jesuit) institution with an undergraduate population of 3617 and student/faculty ratio of 18:1. Of the Dana schools, the Canisius student population is among the most needy. Ninety percent of the Canisius students demonstrate need; 89 percent get aid from Canisius; 43 percent of Canisius families fall below the U.S. median income for a family of four; and 14 percent of Canisius families are at or below the poverty level. Among the Dana initiatives, the Canisius program was the one aimed at the least affluent population. Canisius also has the highest percent of nontraditional age undergraduates (25 percent), which, if those students are considered "independent " by financial aid standards, may contribute to the overall high need of the undergraduate population. Similar to the population it serves, Canisius College is not wealthy, with a modest endowment of $35 million.

To serve this needy population and the needs of the college, Canisius initiated the Dana Earning Excellence Program (DEEP) to offer educationally relevant work experiences that would attract and bond to the College talented students in need of financial assistance. These students might otherwise be forced to choose the lower cost alternative of a public higher education institution or to spend four years in unrewarding routine work to pay for their education. This focus on recruitment is similar to programs at Bucknell and Cornell.

Perhaps the most staggering outcome was a 95 percent retention rate to graduation among DEEP program participants compared to a 50 percent typical four-year cohort retention-to-graduation rate. Although fewer fellowships have been awarded since the termination of Dana funding, the original goals continue to be supported, and the program itself remains unchanged. DEEP set goals to recruit top students, reduce their level of debt, and increase faculty productivity. The program was successful and Canisius has continued its operation.

Selection was based on academic qualifications and aimed at particularly high-potential students who were "at risk" of not enrolling, of not realizing their full potential, or of dropping out.

Housed in the Office of the Academic Vice President, DEEP made internship opportunities known to students through formal announcements in the College newspaper and mailings to those with the GPA required for eligibility. Incoming freshmen were recruited to the program through a personalized letter of invitation. Students identified their work experience from placements that already existed, developed their own, or occasionally were placed by the program. The on-campus projects were developed by faculty and students together, whereas off-campus projects were developed by students. Each year students reapplied and competed for placement based on GPA and appropriateness for position. Participants were not required to write a formal report. Evaluation of the program was planned to measure the five main objectives of increasing the recruitment of talented students, improving retention, enhancing the baccalaureate experience, stimulating faculty scholarship, and increasing financial aid support from external resources to reduce indebtedness. For students who completed internships, public announcements of recognition were posted on bulletin boards, in the student newspaper, and in the alumni newspaper.

The program during the Dana years of support involved between

65 and 90 students a year with a relatively even distribution of freshmen through seniors. Ninety-five percent of the opportunities were actually on campus. Sixty faculty members were involved over the course of Dana funding. Students engaged in research did such things as library research and bibliographic work, assisted in laboratories, edited, proofread, and ran subjects through experiments. Other students engaged in survey research, data processing, and institutional research. Through these collaborative experiences, Canisius faculty began to see their top undergraduates as possible contributors to their professional activities. Faculty publications increased as a result of having a Dana Intern.

Canisius did not involve alumni, as most programs were on campus. However, alumni do support the program financially and the number of named scholarships has increased as a result. Corporations and foundations were involved in supporting the program financially but not through job sponsorship. The Howard Hughes Medical Institute, for example, provided funds for freshman and sophomore science students. The Arthur Vining Davis Foundation continued funding DEEP at the conclusion of the Dana program.

On an annual basis 400 students have been invited to participate in DEEP; 125 apply; and 50 grants are provided—40 to continuing and 10 to new students. Over the four years of the program, 60 freshmen enrolled with a yield of 50 percent, versus a 35 percent yield overall, demonstrating the program's effect on recruitment. Originally the program began with a 3.0 GPA requirement. This was eventually raised to 3.15. Canisius reports that the average GPA of participants has been 3.62. While there was some debt reduction in the first two years of the DEEP program, that aspect was phased out in year three. Finally, faculty collaboration was reported as strong throughout.

Davidson College, 1987

Davidson College is a highly selective private co-educational liberal arts college affiliated with the Presbyterian Church (U.S.A.). Located about 20 miles north of Charlotte, North Carolina, Davidson enrolls approximately 1600 students. Since its establishment in 1837, the College has graduated 21 Rhodes Scholars and is constantly ranked among the top liberal arts colleges in the country by *U.S. News and World Report.*

Davidson's proposal to the Dana Foundation was designed to serve students whose financial need was between 75 percent and 100 percent of the cost of education. The educational experience of higher-need students, Davidson noted, is radically different from those who can afford to pay all or most of their expenses. They found this group to be frequently dominated by racial/ethnic minorities. For these students the four years of undergraduate education are often devoid of special experiences, encounters, and associations which make higher learning a truly liberating experience. The complexity of being a racial or ethnic minority student in a predominately white environment, coupled with the inability to participate in many of the educational opportunities, contributes to the difficulty of recruiting and retaining those students.

Davidson proposed that the Dana Scholars program would offer racial/ethnic minority students who had high or full need the opportunity to participate in academic year and summer work experiences of three types: academic year and summer associates in the "Love of Learning" program for racial/ethnic minorities in which the associates would work with faculty and staff and accomplish the objectives of the program; faculty research assistants who collaborate with faculty on research projects and assist with group arrangements for travel and study abroad; and, finally, summer service interns in both foreign as well as Charlotte-based non-profit organizations. These interns would work on a variety of service projects.

Davidson's program served only about 6 percent of the student body or a total of 95 experiences over the four-year grant period. Seventy-five percent of the experiences took place off campus during the summer. Since the program focused on summer experiences, students could participate a total of three years beginning the summer after freshman year.

Eligible students were made aware of the program through letters and brochures. The only criteria beyond 75-100 percent financial need was to be "in good standing." Students selected placements from those that already existed, some were matched by the programs, while others were recruited by faculty. A resume and application were required for all placements. Faculty selection to participate was based on how they would use student assistance. Off-campus employers, which were the majority, were recruited through Offices of Service Coordinator and Experiential Programs, and friends of the College. Off-campus employers of summer service interns did not receive a wage subsidy by Davidson.

Approximately 15 Dana scholars worked as associates in the "Love of Learning" program spanning one academic year and a summer. Five Dana scholars served as faculty research assistants and between 10 and 25 students worked as summer service interns. Each Dana scholar received an international experience, three summer internships, and three academic year experiences. Davidson did report that they did not use the program as a recruiting tool in any significant way. There was no alumni participation, no data was available regarding increased alumni giving, and no data on debt reduction. Davidson did, however, realize significant success in fundraising—$367,000 from foundations, $5500 from corporations, and $177,000 from alumni.

The program was housed in the Dean's Office for the duration of the grant and then moved to a faculty member. Each of the four years of Dana funding, the program was administered by two part-time staff members.

Davidson reports that the Dana-funded program continues; however, the summer internship has been reduced in size since the Dana grant expired. The "Love of Learning" counselor/mentor support was expanded through other sources and a major new program with 90 summer service interns a year, 30 awarded to freshmen, has been established. Interestingly, Davidson did not report minority statistics as part of its institutional survey, thus it is difficult to measure the success of its original goal regarding the increased participation of racial/ethnic minorities. The program in modified form continues today—with funding from foundations and other sources.

DePauw University, 1987

Located in the town of Greencastle, Indiana, DePauw University is a small liberal arts institution affiliated with the Methodist Church. The approximately 2000 undergraduates enjoy a student/faculty ratio of 12:1. DePauw has a modest endowment of $140 million. The DePauw undergraduate experience features a month-long intercession in which students pursue special projects. About 700 students participate off campus each year.

Because DePauw is almost identical in size to Dickinson, we have taken the liberty to run some comparisons. For example, twice as many internships took place at DePauw. Both programs focused primarily but not exclusively on summer experiences. Students at both Dickinson and DePauw were not guaranteed participation for more than one year; however, Dickinson noted that students often participated more than once. DePauw intentionally wanted to make the University available to all qualified students regardless of their financial background and in doing so tried to maximize participation in the Dana program.

DePauw's Charles A. Dana Apprenticeship Program had three broad components: summer jobs/job internships, research assistantships, and tutorial positions. Within these three categories, the students selected to participate in this program chose from on-campus or off-campus, summer or term-time, and for-profit or not-for-profit employment opportunities.

The criterion for participation was a financial need greater than $4000. The administration of the program was housed in the Financial Aid Office. Day-to-day administration of the program was originally designed to be run by two Dana grant recipients; however, the institutional survey report in 1993 stated that there was one full-time administrator and one student worker. Off-campus employers did not receive wage subsidies from DePauw for their interns. Eligible students were recruited for the Dana program through a direct mailing early in the spring semester. Students identified their work experiences from placements that already existed on campus or developed their own off-campus experiences. A student's GPA was *not* part of the selection criteria. Faculty selected students for their projects based on the strength of their abilities as seen through a writing sample, resume, and personal knowledge of the student. Non-research internship proposals were judged on their viability. The Apprenticeship program was available mostly to upperclass students and provided two opportunities that were previously unavailable to needy students: The Management Fellowship program at the McDermond Center for Management and Entrepreneurship and international business experiences overseas. Internships at the Center were both in business as well as the public sector. There were no efforts made by the University to recognize students who had completed the internship as was the case at a number of the Dana institutions. Nor were there any spirit building activities as part of the program to bring participants together.

During the course of the Dana grant, alumni giving increased by $490,000. In 1988-89, for example, the number of donors increased by 18 percent in response to specific fundraising solicitation for the Dana program. There was no significant corporate involvement. There was a loan reduction component built into the program in year three

(1989). Students involved in summer research committed to 400 hours of work for $3500 and received a $750 loan reduction. Other summer internship participants committed to 480 hours of work and received a $1000 loan reduction.

DePauw noted that the Dana program per se was terminated, but that successor projects including the Science Research Fellows program with grants from the Lily Foundation, Pew Charitable Trust, and Howard Hughes Medical Institute were all modeled after the Dana experience. However, before the program changed, DePauw cited three benefits to the institution as a result of the Dana grant: it aided faculty in research; attracted high-quality students as tutors; and created good relationships with industry and not-for-profit organizations.

Franklin & Marshall College, 1987

Franklin & Marshall College is located in the city of Lancaster, Pennsylvania, with 1900 undergraduates. It boasts an endowment of $153 million. Philadelphia, 130 miles away, is the closest urban center. This small, liberal arts college also offers a pre-professional program in business. The overall student/faculty ratio is 11:1 and there are slightly more male than female students.

Franklin & Marshall's Dana-funded program focused primarily on off-campus, summer employment which is similar to the program at the University of Rochester (SummerReach). Both programs emphasized pre-professional employment opportunities as a tool to assist undergraduates with making career choices. Both programs subsidized employers.

Franklin & Marshall's proposal created the Career Exploration Internship program (CEI). The program goal called for 100 rising sophomores, juniors, and seniors to participate with paid term time and summer internships primarily outside the College community. The internships were to provide a comprehensive and useful introduction to a variety of professional career choices while at the same time to introduce an additional, meaningful way to help deserving students pay for their educational costs. The 100 upperclass internships were proposed to be distributed into 20 academic year opportunities at 10 hours a week for 35 weeks and 80 summer time service interns off campus for 10 weeks at 35 hours a week. However, the institutional survey indicated that summer internships were evenly split between on- and off-campus placements. Stipends of $1500 were provided. The 1990 interim progress report, however, indicated that the average stipend was increased from $1500 to $2700.

Criteria for selection included academic promise, extracurricular potential, recommendations or special achievement, and the need for financial assistance. Because Franklin & Marshall gapped financial aid students by not meeting their full need, debt reduction was a fixed component of the CEI program. CEI received four applications for every internship position. Almost 95 percent of the participants were juniors and seniors. Program participants were fairly evenly distributed over the four-year grant period.

The program was housed in the Career Services Office and administered by one full-time staff person. Students developed their own experiences. Students were notified about the program through newsletters and meetings and applied by submitting a resume and writing sample; interviews with sponsors were also required. This process of finding a summer job is similar to that at Rochester in that the student learns valuable life skills in this pursuit of a position. In addition, some sponsors used GPA and coursework as selection criteria. Off-campus employers were recruited through personal visits and mailings. Franklin & Marshall drew alumni already involved with the College to supervise as well as develop internships. The program generated new corporate internships and new corporate support.

Examples of three typical on-campus experiences for the CEI program were: working in the Office of Planning and Institutional Research gathering data, aiding and interpreting that data to support decision-making policies of the College, managing an informational data base, etc.; working in the Development Office to assist the director of annual giving in organizing and implementing the senior class gift project; and assisting in the Chemistry Department in the research on special acid-base interactions.

Off-campus work included such placements as: the Pennsylvania Dutch Convention and Visitors' Bureau—assisting the Director of Marketing, helping the various market research projects with promotional brochures and aiding with public relations activities; the Hebrew Rehabilitation Center for the Aged—participating in analysis of several gerontological studies; and an assistant's position at GTE Telecom Inc.—supporting in a wide variety of accounting projects and functions.

When Dana funding ended, employer subsidies at Franklin &

Marshall and board stipends ended. At present, the College relies totally on sponsors who are willing to pay all costs for students and are able to absorb some of the housing costs for summer employment.

Smith College, 1987

Located in the small town of Northampton, Massachusetts, Smith College is one of three all-women's colleges funded by Dana. (The other Dana-funded women's colleges are Bryn Mawr in Pennsylvania and Barnard in New York.) With approximately 2900 undergraduates the student/faculty ratio is 10:1. As a member of the Five College Consortium with Amherst, Hampshire, Mount Holyoke, and the University of Massachusetts, many academic, social, and cultural resources are offered through joint faculty appointments, shared facilities, and other cooperative arrangements. Smith has an endowment of $435 million.

Smith established a program of Dana Student Interns which concentrated on two areas. The first provided on-campus and off-campus academically-oriented summer internships to financial aid students; the second provided funds for teaching assistantships in elementary and secondary schools. Smith was concerned because its graduates were leaving the institution with between $8000 and $9000 indebtedness. Smith speculated that their alumnae were choosing first career opportunities for financial rather than developmental reasons. Smith also wanted to get more of its alumnae into the teaching ranks and into science-related occupations. The goals, therefore, of the Smith Dana Student Internship program were: to reduce indebtedness so people choose careers for the right reasons and make good decisions; and to ensure that Smith was accessible as a private institution to students of all economic backgrounds; and to provide access to educationally purposeful summer work for financially needy students.

The majority of the programs were summer and primarily off-campus. Academic year opportunities were on campus as expected. First year students did not participate in these programs and the majority were juniors and seniors, all of whom demonstrated financial need. Of note is that 17 percent of the participants were Ada Comstock, non-traditional students specially recruited to Smith College. Average pay was just under $2000, with virtually no course credit. Promotion of the Dana program did not appear in admissions literature. However, Smith admissions officers specifically talked of the Dana internship program to students, parents, guidance counselors, etc. The admissions video features the Director of Career Development describing the Dana internship opportunities. There was no measurable recruitment impact reported.

Students were made aware of the program through informational meetings, announcements in publications, and through the Career Development staff. Students found their work experiences through placements that already existed and developed their own, sometimes in collaboration with a faculty member. Selection criteria for students did not include GPA. However, students did have to submit a written proposal that was judged on how well the internship related to the student's academic goals. Letters from faculty and internship sponsors were also considered. For on-campus placement any appropriate faculty member engaged in research could offer to supervise an intern. Off-campus employers were developed by students themselves, alumni, and the Career Development Office. Students could participate only once and a formal written report was required upon completion.

Research activities with faculty included performing laboratory techniques, statistical analyses, archival research, annotated bibliographies, design of experiments in the social sciences, etc. Teaching assistants primarily involved student teaching for education majors at the Smith campus school. Faculty publications emanating from the Dana program included self-published "Notes and Readings for Summer Program in Arts Management"; "Public Opinion Toward U.S. Involvement in Central America"; an American Political Science Association conference paper, Chicago, Illinois, September 1987; "Toward a Theory of Trade and Services—Prospects for Service Exports in Developing Countries" submitted for publication; and other unpublished manuscripts. The *Smith News*, fall 1989 edition, featured a story on the "Dazzling Danas," including a member of the Class of 1991 who began her internship at the Williamstown Theater Festival, another member of that class who studied circadian physiology—that is, biological changes governed by internal time clocks—and a California native of the Class of 1991 whose internship was with the Congressional Caucus for Women's Issues in Washington, D.C.

Smith also joined the ranks of Bryn Mawr and others noting the labor intensity of the administrative aspects stewarding the Dana program. The Smith program was administered jointly by the Office of the Associate Dean of Faculty and the Career Development Office; staffing equalled one part-time person.

Faculty productivity was increased and the program helped establish a collaborative attitude between students and faculty. Although Dana teaching fellows were scheduled to receive a $1500 forgivable loan for those who taught up to four years, there was no evidence of that in the 1990 program report or the 1993 institutional survey. Financial awards were given in lump sum stipends in ranges from $270 up to $4745. Smith was successful in its fundraising, having received $70,000 from four foundations, $23,000 from three corporations, and $216,000 from parents, alumni, and friends.

Alumnae were involved in both supervising and developing internships. The Career Development Office mailed letters to alumnae and Smith Clubs with information about the Dana Grant and requested their assistance with internships. Alumnae were also asked for contributions in the form of matching gifts. An endowment was established through alumnae and corporate gifts. Although the Dana funding has ended at Smith, the program now falls under the umbrella of the Summer Internship Program and shares the original Dana goals. Smith reported that other internships in the sciences were attracted because of the Dana program and new foundation support funded by Sherman Fairchild and the Schultz Foundation were received as a result of Dana.

Union College, 1987

Located in the city of Schenectady, New York, near Albany, Union College is a small liberal arts college which also offers a pre-professional program in engineering. Enrollment is about 2000 undergraduates with approximately 45 percent receiving financial aid. The student/faculty ratio is 9:1 and the endowment is modest at almost $120 million.

Union shared the most basic goals of the Dana SAEQ series of grants. They proposed providing work experiences relevant to academic work and to career development with an eye on debt reduction. Dana funds also helped to expand the College Work Study Program. All work opportunities were on campus, which makes Union's program similar to Birmingham-Southern, Bates, and Franklin & Marshall's. Other programs that were exclusively on campus include those at Bucknell, and Canisius, but their programs served students all four years, were marketed more aggressively to incoming freshmen, and measured greater levels of debt reduction. This is not to say that Union did not focus on debt reduction, but that when institutions involved students all four years, debt reduction was usually a more significant outcome. In addition, Union's program was somewhat similar to Vassar's in its use of Dana funding to expand College Work Study funds.

Union already was providing a "Horizons Program," a summer experience for science and engineering students, and saw the Dana grant as a way to expand on that model. This program complemented a social science research program launched the year before Dana funding began. The request to Dana called for 65 students a year or 260 over four years to be split between the summer research and academic year teaching assistanceships in both of the above programs. At the end of the Dana grant period, 319 total student placements had actually occurred, mostly filled by third or fourth year students. Some students participated more than once, thus the actual number of students involved was less.

At the end of the grant, the academic year program was dropped, but the summer research opportunities continue today with some foundation and corporate support. The program, while in existence, was reported to support more work study positions during the academic year and strengthen the summer research program. Conversely, the academic year opportunity offered more relevant educationally purposeful work.

The program was coordinated by a faculty member who publicized the program through the student newspaper and through other faculty members. Students were selected based on faculty recommendations. Faculty were eligible based on their willingness to sponsor a student. Approximately 75 faculty members were involved during the grant period.

To recognize completion of internships, students were required to file a formal report on their research project. Typical summer research positions required a faculty member sponsor. Academic year teaching assistants were chosen by the program director, based on qualifications including experience, preparation, interest in teaching as a career, etc. For those few freshmen who were recruited directly into the summer research program, a press release was sent to the student's hometown newspaper. Stipends for the summer program were $1250 plus room. Teaching assistants during the academic year received $1500 based on working 10 hours per week for 30 weeks at $5.00 per hour.

Since the Union program was predominantly on campus, where students worked with faculty members, alumni participation was minimal. Some new corporation and foundation support for the summer work opportunities was generated as a result of Dana funding. Information was not available regarding exact debt reduction outcomes. The summer research program is now College funded with a modest amount of foundation and corporate support.

Vassar College, 1987

Located in Poughkeepsie, New York, Vassar College is approximately 70 miles from New York City. Founded as a college for women, Vassar has been coeducational since 1969. Today, men comprise 42 percent of the student body. The student/faculty ratio is 10.5:1. A Vassar education is steeped in the liberal arts tradition, but there is no required core curriculum. Students may choose from four different paths leading to a rigorous academic major. Most faculty live on or near the campus; a number of faculty also serve as "house fellows" in the residence halls, further enriching the contact between students and their professors. Vassar's tuition ranked second among all Dana-funded schools. Its endowment of $270 million ranked 60th among all colleges and universities in the United States.

Vassar proposed that the Dana grant help them to extend, diversify, and improve upon two existing academically-oriented student employment programs. The first involved students during the regular academic term in a longstanding effort called the Departmental Internship Program or "junior colleagues." Students were selected on the basis of specific abilities. Vassar proposed to increase the number of these interns. The second program was a summer program, a new formalization of existing ad hoc arrangements which had already been planned for the summer of 1986 for the academic departments in the areas of mathematics and science. The initiative was called the Undergraduate Research Summer Institute (URSI). In the Institute, students worked directly with faculty members as research assistants at a relatively high level of skill and responsibility. Vassar proposed to fund this program with the Dana grant in order to offer assistance to faculty members in all academic departments. It is of interest to note from the participants' responses that in expanding two existing programs the recipients of the Vassar Dana grant were unaware of SAEQ and the source of the funds.

The plan called for 80-90 students a year to participate. Vassar's program is most similar to that at Oberlin in that both sought to increase student/faculty interaction and both used the Dana funds to expand existing programs and to creatively utilize College Work Study funds.

A total of 316 Dana interns were involved in the programs from academic year 1987 through 1992. Seventy percent of all the internships took place during the academic year, all were on campus. The 30 percent which took place during the summer were heavily off-campus (85 percent). The program itself was housed in the Dean's Office and administered by one part-time staff person and two student workers.

Of the two existing programs, the "junior colleague" program was a longstanding tradition at Vassar. Duties included being a lab assistant, tutoring, grading exams, assisting faculty with projects, coaching language labs, etc. This program historically was not need-based. The Undergraduate Research Summer Institute (URSI), although not as well established as the "junior colleagues" program, allowed students and faculty to work together as co-workers and co-authors. The original grant for this program came from the National Science Foundation. Vassar sought to increase the number of interns from 60 to 100. The additional interns were entirely from the needy student population. Vassar further planned to use the existing procedures to conduct the programs while expanding opportunities to administrative departments previously not eligible, such as the library and the art gallery. By its nature there was virtually no alumni involvement nor corporation and foundation partnership although there was new financial support sought and gained from foundations and corporations. Vassar listed $50,000 in foundation support, $25,000 from corporations, and $213,000 from alumni. Vassar did not note any success with debt reduction nor were these internships used as a recruitment tool.

Looking to the future, Vassar reports that Dana internships have merged with two other summer research internship programs. The academic year internships continue as they have for more than 50 years. The independent summer projects activity was terminated. However, the Dana program helped establish a summer research internship model and it relieved the College's budget for campus work for six years. It also helped to clarify priorities for student support other than term time scholarship.

Wheaton College, 1987

Wheaton College has one of the smallest student population among the Dana funded institutions at 1300 students. Located in Norton, Massachusetts, it is half-way between Boston and Providence in a small-town, suburban setting. Despite its small size, its endowment is $58 million. The institution was founded in 1834 as a women's college with a mission to prepare its graduates for challenges and choices in a changing world. In 1988, Wheaton became co-ed. It is still 65 percent female. The timing of the change from the single-sex institution and the award of the Dana grant was coincidentally beneficial to the College. The Dana funding provided an opportunity to build on the mission of preparing graduates for the changing world by creating an institutional culture around work—a gender neutral issue. The Dana Program arrived on campus at the exact time the College was seeking an institutional signature.

Wheaton's proposal was distinctive. The main feature of the Wheaton Dana Fellows Program was the emphasis on work as an evaluated learning experience. The Dana Fellows program was the pilot for Wheaton's new Filene's Center for Work and Learning. The Center's goal is to integrate an explicit learning component into all the work-related activities sponsored by the College. This innovative program sought to move beyond the hit-or-miss approach of learning from experience by providing a structured framework in which students could reflect on their work and subsequently document the personal and professional growth results.

The Dana Fellows program awarded 40 of Wheaton's most promising and motivated financial-aid students each year. Like many of these programs, work opportunities for Dana Fellows would include student/faculty collaborations, projects in key administrative offices on campus, and field placements in non-profit organizations and corporate settings in the surrounding community during the academic year or world-wide during the summer.

Following a decline in interest in the January internship program, one goal of the Dana program was to reintroduce and to re-energize students into experiential learning opportunities. The Wheaton program took the Dana proposal and integrated it with other emerging actions, activities, issues, and priorities on campus. Wheaton wanted to re-think learning and work and revise all forms of work from Work Study to internships to field placements (an academic requirement).

Wheaton wanted to ensure that learning can and does take place in many arenas and unite rather than oppose the value of careers and liberal learning. To do this, the College believed it had to make internships and field placements more broadly accessible regardless of financial circumstances. The Dana grant was a test pilot for this model.

Wheaton noted that the work experience alone was no guarantor of learning. One must analyze the work experience and articulate what's been learned. Because the College was convinced that work must be an evaluated learning experience, a good deal of interaction was required between employers and the faculty. The programs in the Center for Work and Learning were multifaceted. They expanded the current bank of 500+ internships and field placements nationwide by creating a larger number of summer and paid opportunities while also providing a forum for individual and group assessment of work experiences. In addition, the integration of an explicit learning component was added to jobs on campus. Outreach efforts were undertaken to establish linkages between alumni and students. In particular a minority network program was begun. Course(s) in the social sciences addressing cultural identity in a multicultural world were established, many with field work. Two visiting lecturers came to campus for three to six days each year. These seminars featured field supervisors and members of the Wheaton Associates, a prestigious group of alumni and friends involved in business and industry. Wheaton faculty also engaged in summer internships/seminars with business leaders to discuss the role of liberal arts in career preparation. Events were regularly scheduled to help students prepare to deal with high level jobs and dual career families.

Results of the Dana Fellows Program demonstrate increased faculty productivity, and increased alumni and corporate involvement with the College. An internship developer in the Center for Work and Learning was assigned to contact alumni, corporations, social service and arts agencies, and government sponsors to develop summer and January term paid internships. The Provost headed up an effort on campus encouraging faculty and administrative offices to develop collaborative learning opportunities. Students were encouraged also to develop their own opportunities. Finally, two Dana awards were reserved for summer overseas internships through the Global Awareness Program.

Like Cornell, Wheaton took special efforts to create a sense of community. Once a semester Dana Fellows got together with faculty facilitators to reflect on work, share experiences, and make presentations. Summer interns first received an orientation about evaluated learning prior to their work experience and at the end of the summer were required to present a paper or project assessing the outcome. In addition, there was a post internship seminar on work and learning.

Through the third year of the program 123 internships were completed, 103 of which were unduplicated. Students were all needy; the experiences were relevant to coursework and pertinent to career and major. Internships paid $2000 to $3000 in the summer and $900 to $1100 during the term. Not surprising, with such a comprehensive approach, Wheaton generated over half a million dollars in financial support in the first three years of the Dana grant. As the program matured the College considered involving all students, not just needy students, because there was some resentment from non-needy students who couldn't participate in these educationally enriching opportunities.

A few experiences stand out in the Wheaton Dana Fellows Program. First, the College reported that 27 institutions inquired about Wheaton's Student Aid for Educational Quality Dana Fellows' activity, and nine institutions actually visited Wheaton College! With the exception of Cornell, no other institution seems to have generated that much peer interest. Second, the presentation of the Fellows Program in the admissions literature was superlative. Finally, Wheaton created a work and public service record transcript that supplements formally a student's academic work record at the College.

Wheaton is clearly an example of an institution where the Dana Fellows Program made a tremendous difference. While the outcome is similar to that at Furman, the driving forces behind each program were different. Wheaton was in search of a signature as it emerged a new coeducational institution, whereas at Furman, the Dana program became the proxy for institutional pride and academic quality.

Barnard, 1988

An all-women's liberal arts college, Barnard College is affiliated with Columbia University, located in New York City. Barnard is one of three all-women's colleges in this research, with an enrollment of 2200. Its student/faculty ratio is 12:1. Its endowment is $56 million.

Because of its New York City location, Barnard offers cooperative degree programs such as a 3-2 degree program with Columbia University's School of International and Public Affairs, and an arrangement of off-campus study with Manhattan School of Music, Albert A. List College of the Jewish Theological Seminary of America, and the Julliard School of Music. The opportunities for off-campus internships were many and varied. More fundamentally, Barnard holds that an "important goal of a liberal arts education is its application in the world at large and the College encourages its students and faculty to participate in work activities and public service, both of which compliment education in the classroom." As one of two institutions funded in 1988, the last year Dana made these grants, the Barnard Dana Internship program proposed to enable needy students who couldn't otherwise combine public service and career development to gain advanced knowledge through research. Opportunities were available in the arts, communication, medical, and non-profit communities of New York City. In addition, a FIPSE grant evolved as a result of the Dana grant, which is discussed at the end of the Dana Program description.

The Dana program was primarily targeted to upperclass students. Less than 3 percent of the participants were first year students. With the majority spread among upperclass students, the highest participation was among seniors. Internships took place both on and off campus, and were exclusively during the academic year. The on-campus experiences tended to focus on research, writing, and presentation while the off-campus internships were involved primarily in not-for-profit community service organizations working in research, interaction with clients, writing and editing, administrative duties, program development, translation, public relations, artistic performance, etc. Average earnings in the internship program were $1200 a year. The range was from $300 to $2500. Only four-fifths of the students, Barnard reported, completed the internships primarily because students experienced problems with scheduling and conflicts between student life and the work world. Although the program was heavily marketed in the admissions literature, Barnard cited that there was no empirical evidence of impact on enrollment. There was a good deal of alumnae participation, supervision, and creating of internships as well as alumnae contributions. The program continues to benefit from $100,000 a year annual support from the College since the expiration of the Dana grant.

The program was administered through the Career Services Office by one part-time worker and one to two student workers. These opportunities generated through the Dana grant complimented experiential positions that already existed. Students were also able to develop their own opportunities. Students were made aware of the program through announcements in Career Services publications and through counselors' and faculty contact with students. Faculty were invited to sponsor an intern through a general mailing and off-campus employers were already part of Career Services' general internship listings. Students were not guaranteed participation for more than one year and interviews with program staff and sponsors were required. A formal report was required at the completion of the internship and a reception was held to honor Dana interns once a year.

Examples of students working in the arts include the New York Youth Symphony, the Drama League of New York, and Orion Pictures. The medical communities such as St. Luke's Roosevelt Hospital were assisted in the emergency room by Barnard students. On the world scale, students worked for Care International researching international development projects such as the Soviet food relief effort and assisted both faculty and non-profit organizations. Diversity of departments served is perhaps best reflected in the third year of the program when 30 on-campus students participated in 16 different departments and programs ranging from Biology to Environmental Studies to Dance and Theater.

A grant from the Fund for the Improvement of Postsecondary Education (FIPSE) studied the effects of the Dana and a similar community service-oriented internship program (Ford) as the choices students made to get involved with not-for-profit service positions. Within the focus on the Dana and Ford interns, the findings confirmed the belief that paying students for their community service does involve students who might otherwise have had to hold paying jobs rather than contribute to their communities if they had to do so on a volunteer basis only. The Dana and Ford grants did enable the majority of the interns not to hold an additional paid, part-time job, and did help in reducing the debt that those students accumulated during college. The Dana and Ford group did not, however, become more service-oriented than their classmates as a result of the internships. Because the Dana/Ford group did not differ significantly on the outcomes when compared to the students whose community service experience had been non-paid, it cannot be asserted that paying community service interns instills a stronger commitment or more positive attitude towards service. It is significant, though, that the financial grants did enable students to participate in service when they might otherwise not have been able to do so.

Because Barnard College was involved with surveying its program alumni for the FIPSE grant, it did not participate in the alumni survey, thus there is no analysis of the Barnard students' experiences.

Lafayette College, 1988

Lafayette College, founded in 1826, is an independent, coeducational, residential, undergraduate institution with a student population of 2045 men and women of high intellectual promise and diverse backgrounds representing 48 states and over 30 foreign countries. The College's curriculum is distinguished by the rare combination, on an undergraduate campus, of degree programs in the liberal arts and in engineering. Lafayette's endowment per student is in the top two percent of all institutions in the country. Lafayette is located in Easton, Pennsylvania, just an hour and a half from both New York City and Philadelphia.

The history of Lafayette College regarding work is special. The first President, George Jurkin, had Trustees amend the College charter to include mechanical and agricultural work as part of the Lafayette experience. In fact, the first building on campus, Lafayette notes, was built with student labor. In that spirit, Lafayette established the Dana Venture program and Dana Venture Scholars with SAEQ funding. The goal of the program was to expand Lafayette's small but successful student research assistantship program, contribute to the writing across the curriculum program, and provide aid to students who would work on educationally meaningful projects.

The program was divided into two parts. The Lafayette Dana Venture Program included two weeks of training and a summer work experience along with an academic internship in the January term—all originally scheduled to focus on involvement outside a student's major. The second, which was never launched successfully, was the creation of an alumni summer job network to guarantee educationally meaningful summer employment to Dana Venture Scholars. The program also aimed at matriculating good students. Lafayette planned to evaluate the program based on changes in yield, debt reduction of $2500, number of alumni involved, and a survey of student participants' and employers' satisfaction with the program.

The program was housed in the Provost's Office. Faculty submitted proposals for research and outlined the student's role and often identified the student they wished to hire. Students could also request to work with a specific faculty member. Otherwise students were recruited for the program through positing of the various opportunities. While students from all class years could participate, the majority were sophomores and juniors. Participants were not guaranteed a second experience, but continuation projects were given priority whether they wanted to continue with the same faculty member or a different one. Faculty were paid $500 for supervising Dana scholars. When the internship ended, students were not required to write a final report.

In the final analysis, no alumni were involved. No information regarding debt reduction was available. Although Lafayette reported that the Dana program had no impact on campus, as many as 40 percent of the research assistants listed as participants in the Dana program were students from foreign countries. Twenty-one Lafayette College students gave papers over the course of the Dana grant at national conferences on undergraduate research. In the future, Lafayette desires to provide a capstone experience for students to present their research findings at the end of their summer experience. The Lafayette College program is one that matured and changed over the course of the Dana grant and emerged as one not focused on needy students or debt reduction but on providing students with enhanced learning opportunities through collaboration and research experiences between faculty and students. Ultimately the program came to be known as the Excel program, a student research effort based on the Dana Model. The Excel program today is not limited to students with need. Today the summer board stipend awarded in the Dana program has been removed.

The outcomes at Lafayette are similar to those at Birmingham-Southern in two ways. Both proposed off-campus summer experiences with limited results. (Even though Birmingham-Southern actually launched its summer program it never reached the numbers proposed and was abandoned after Dana funding ran out.) In addition, both cited that the benefits to their institutions were improved interaction between faculty and students and that faculty viewed motivated students in a new, prestigious light.

Lafayette College did not participate in the alumni survey. As a result there will be no data available from Lafayette College alumni on their experiences in the Dana Venture Scholars program.

Chapter 4

What Works, What Doesn't, What Makes a Difference

The overwhelming response from the participants of these career-related, educationally purposeful, work experiences spanning 20 different campuses and programs as well as the institutional data collected from 23 participating schools, provides a very rich, fertile ground for analysis. *Appendices II and III* cover such issues as response rates by institution, research design and methodology used in both univariate and multivariate analysis with definitions of statistical significance, and special statistical segmentation packages such as Chi-square Automatic Interactive Detector (CHAID). In telling this story of the outcomes and benefits of these experiences, the data have been analyzed in four clusters of groupings. First, all respondents segmented by Cornell, Dana SAEQ schools, Duke, and Rochester. Second, research results of Cornell-only respondents. Third, research results of the Dana, Duke, and Rochester participants on selected variables. And, finally, an examination of the differences and experiences for women who attended single-sex institutions (Bryn Mawr and Smith versus the other Dana institutions.

In the first of these four analyses there is a good deal of program description information included, such as what was the type of placement and what effect did that have on the experience? Was on-campus more or less rewarding? Did the off-campus experience have certain attributes and benefits? What happened when the support for the intern was lacking? What about the make-up of the experience itself? With whom was the student working and what were the responsibilities?

The next set of data examine the type of experience the student participated in, again in the aggregate as well as divided by the four institutional groupings: Cornell, Dana SAEQ, Duke, and Rochester.

Questions include the significance of whether the experience was related to the major, if it was during the academic year how many hours were involved, and did that play a role as a positive experience or was it a detriment? What benefits did the type of experience have in preparing one for the work place? Was one experience sufficient or did benefits accumulate when one student was placed in multiple settings? Did it seem to make a difference whether the experience was over the summer when the press of the undergraduate program wasn't part of the mix or did an experience over the academic year bring with it advantages or barriers?

A good deal of the information sought from participants had to do with developmental outcomes, be they academic or personal. In particular, such questions arose as to whether the experience re-affirmed or redirected a student's intended area of academic concentration? Did it influence their plans after graduation? Did it help or hinder their academic performance while in school? Did it provide any additional, supplemental experience to the curriculum which otherwise would not have been available? On the personal side it was a question of development of skills and values. Did participants have a sense of commitment or obligation to others? Did they have a stronger identification with citizenship and responsibility? What skills, if any, were developed as a result of these experiences including such things as communications skills, team work, self confidence, etc.? Which acquired skills, talents, and values link up with certain experiences and seem to make the biggest difference? Finally, in this first very general sweep of the results of these experiences, information regarding institutional benefits including the influence on recruitment, the participant's relationship to alma mater, and the individual's financial benefits for participating in the program are examined.

The second section of the analysis focuses only on Cornell because of the rich history and longitudinal data available and of course because more than a third of all respondents were themselves members of The Cornell Tradition program. In this section the Cornell participant is profiled and three analyses of the data specific to the Cornell program are analyzed, namely 1) the time of the placement, summer or academic year; 2) whether the respondent participated exclusively in the original program or the modified activities: was the participant involved with the Tradition program in 1987 or before but not after that; versus a Tradition participant in 1988 or after but not before. (The previous chapter describes in detail the differences between the two programs.) The changes made were compared for the different results they produced. And, finally, 3) an analysis is done of the Cornell experience attributes and personal developments with the outcomes realized.

In the third section, the experience, attributes, and personal development outcomes are studied for all respondents (other than those from Cornell), alumni of the Dana SAEQ programs, Duke, and Rochester.

And, finally, because Smith and Bryn Mawr were two of the 17 Dana participating institutions in the alumni survey, it seemed appropriate to measure whether these experiences produced different outcomes in a single-sex environment versus those at coeducational institutions. As is true throughout the study, only those differences that are statistically different are reported out. You can assume on all other measures the Smith and Bryn Mawr graduate did not differ in any important way in responding than those women from the other Dana institutions.

Throughout this chapter there will be an attempt to tie the programmatic distinctives as noted in the prior chapter with the particular outcomes. In some cases, because the experience is somewhat widespread (e.g., research assistantship with a faculty member, off-campus experience with an alumnus during the summer in a for-profit organization), it serves little purpose to tie it to a particular institution or program. Certainly thorough reading and referencing of the prior chapter will be important in order to make the unstated linkages and garner a full sense of these outcomes.

I. Type of Placement (See Table 4-1)

Clearly the research assistantship with a faculty member on campus was the most frequently cited type of placement. This opportunity was dominated by the Dana schools, many of whom used the SAEQ program to increase faculty productivity and as a faculty development/recruitment tool as well as a vehicle to enhance the quality of the undergraduate experience. The administrative internship on campus is really a generic "catch-all" descriptor. The placements which students cited vary from institution to institution and even within institution. There were many exciting opportunities described, including

Table 4-1: Type of Placement

	All	Cornell	Dana	Duke	Rochester
Research assistantship with a faculty member	31%	21%	41%	11%	19%
Teaching assistantship with a faculty member	11%	11%	13%	—	—
Administrative internship on campus	29%	53%	18%	4%	7%
Internship with a public service, not-for-profit organization	26%	30%	21%	22%	46%
Internship with a government agency	6%	7%	5%	8%	15%
Internship with a for-profit organization	19%	19%	13%	58%	27%

working in public relations offices, development and alumni relations offices, student affairs, career services, and admissions.

> "I received money from the Dana Foundation to work at a school investigating the effects of exposure to light on human sleeping cycles (circadian rhythms). It was one of the most challenging jobs I have had. It required that I work with other technicians on a randomly generated shift schedule so that the live-in subject would never figure out what time of day it was. For three months I went without a regular sleeping pattern. The job was fairly difficult because of the complexity of the lab and the demanding and competitive attitude of the researcher. It was, in any event, extremely interesting. In all subsequent job interviews, it has raised the most interest and inquiry. Apart from learning a lot about sleep, I learned that I did not want to devote my life to clinical psychological research; something I had been thinking about. This was valuable in helping me determine a career path. I have always been thankful to the Dana Foundation grant, because as a financial aid student, I would never have had a chance to work in a job that would allow me to explore career options and begin to develop a base of interesting work experience to help in future job searches."

> "How Furman and I used the Dana Foundation money may have been a bit unusual. It may be of benefit to explain it because it was successful and helped many people, including me. I worked in one of the offices. Part of my salary was paid via Work/Study, part from the University, and part from the Dana Foundation. During that time, I led a group of 20 volunteer students. Our mission was to encourage, assist, and find career-related work experiences for students in the school. How did it help me personally? Through my work, I received an internship in marketing and computers. I can't remember if Dana money was used, but Dana money did help administratively to get that job. The three-month experience made my resume stand out when I eventually interviewed with a company, which helped get my job today. It is a dream job, with a foundation established from my internship, and the Dana Foundation. Anything we can do to encourage kids to spend some time getting career-related work experiences, we should do. It is the right thing to do, and it works. I am proof of that."

The other end of this spectrum involves a large number of menial, routine assignments, the archetypal experience of which is working in food services. As can be seen from table 4-1, Cornell dominates this type of placement. The Cornell alumni reported that a large number of those placements did in fact take place in the dining halls. As one would expect, the quality of the experience is reflected.

> "It's great to reward students who work but many complained about the program requirements. Working and volunteering does interfere with coursework and the program discourages going abroad—in my opinion, an even more valuable experience than working. Especially if you're working in a dining hall!"

Other things such as debt reduction and participating in The Cornell Tradition community then take on value if the experience itself is to be viewed as positive.

To round out the on-campus experience, there was a marked difference between a relatively small number of students who were teaching assistants (11 percent overall) and how different that experience was described in relationship to the faculty research assistantships where there seemed to be much more mentoring taking place. In fact, faculty were most influential in a research assistantship environment, encouraging, for example, participation in graduate education. The same sense of being mentored was not reported among those who were teaching assistants.

By contrast, the off-campus experiences were of basically two types. The majority of students participated in public-service, not-for-profit organizational settings and when combined with those who were placed with government agencies, almost one out of every three program respondents, or 900 students, enjoyed this type of experience. The linkages here were strong. The not-for-profit sector provided more life-shaping experiences and the value of civic responsibility was closely linked as an outcome.

> "The Dana program has had a substantial impact on my life. I worked for an education lobbying organization and learned a tremendous amount which I have been able to apply in my life both academically and professionally. My internship helped me to develop and produce my Senior Honors Thesis, and to come to the decision to get a Ph.D. in political science (with a major in public policy). As well, my internship led, in part, to my decision to take a year off after graduating from Wesleyan and work 'hands on' in the education field. I worked at a residential treatment center for juvenile delinquents and adolescents in foster care from New York City.

I owe a lot of my current satisfaction and my goals to the opportunities provided by the Dana grant. Thank you!"

"I didn't participate in the Cornell Tradition jobs program. I received a fellowship. During my junior and senior years, however, I participated in a similar program with CIVITAS (a Cornell sponsored community service agency). Cornell paid 90% of my wages while I worked for the City of Ithaca. During my sophomore year I applied for a summer job with the new Cornell Tradition. I got no offers. In part, I believe, this was due to the limited number of cities I said I'd be willing to work in.

One of the reasons I only put two cities on my application was that I was 19 and not really aware of how to go somewhere new and find an apartment for myself. The other reason was money. I knew the cost of living at college or with my parents. Living anywhere else would have to mean that I'd have to earn enough money not only to offset my expenses, but also to pay for school in the fall. I couldn't afford to take any chances. Instead, I spent the summer waiting tables, mowing lawns and feeding lab rats.

Before fall semester started I received a letter informing me I had received a fellowship. During the week before I received the letter, I had been worrying about how I could meet the next tuition increase. I had decided to try one more semester, then transfer. The fellowship not only assured that I could stay at Cornell, but gave me the security I needed to find the CIVITAS job, rather than take the first offer I could get.

The supervisor on the internship was constantly preparing me for the work I would face 'someday when you're heading up a multi-million dollar company.' Working there fostered my interest in the public sector. Being relatively free of debt allowed me to express the interest by joining the Peace Corps. Since the Peace Corps, I've worked for the USDA, run a non-profit corporation, and gone to graduate school. If the success of a career is measured in money, mine has been dismal so far. I measure it, however, in working for what I believe in. I think it's been a success. The Cornell Tradition gave me the financial basis and the encouragement I needed to do what I thought was right.

By the way, the non-profit corporation I ran owned $3 million in assets. I used every bit of advice my supervisor gave me years ago."

More will be said about development of civic responsibility because it is one of the benefits which matches up with the largest number of positively viewed attributes. When a participant reported that as a result of the experience they developed a greater sense of civic responsibility, they were most likely to attach the largest number of positive attributes gained, or skills acquired.

Work in the for-profit sector was also well received by almost 20 percent of those who responded, well over 500 students. This experience linked up most frequently with gaining professional contacts, more of which will be discussed later under Experience, Attributes, and Personal Development.

This analysis wouldn't be complete if it didn't talk about what went wrong. As one might expect when the experiences of over 2500 students are examined, there are many misfires, miscues, misunderstandings, and miscommunications. Poor support for the activities actually broke down into two distinct categories: 1) bad fit, that is the process of identifying individuals and matching them to sponsors and agencies, and 2) not delivering what was promised, when things didn't work as planned. Regarding the former, as noted by this participant,

> "The program was very young when I participated, and it did not have a broad enough contact base at that time. When I entered the program, I requested leads that either involved museum work or a job in the advertising business. I was contemplating an art history major, but wanted to see what jobs were available to me in this field and if I could survive financially if I chose this career path. While I was in the searching stage (which lasted over five months) I did not get much direction in my field. It was my perception, that there was more emphasis on more popular fields such as business and engineering internships. In desperation, I took a job with a company in my hometown that organizes conventions because they had an art department."

The second kind of failure was poor support or direction from the sponsor, that is students were set off on a course with no help, guidance, or assistance, given responsibilities that they were not equipped to perform, expected to produce results that were not commensurate with their background and knowledge, or in some cases just not appreciated. There were not legions of examples of this type of outcome; however, the anecdotal evidence is pointed and poignant.

> "Unfortunately, the faculty member in the biochemistry department I did my project with was very negative in many ways. He always made me feel inferior and was very hesitant to listen to any comments I would have. His research method was do as he said and often went down many dead ends (I realize this is normal in research). However, when he

personally made a mistake he would blame it on myself or another worker when speaking with his colleagues."

"(The faculty member) expected us to work on the level of graduate students when we had never been in a research lab situation. Then he would criticize us for it because we hadn't made enough progress. He gave us no protocols, only original papers and left it up to us to figure out on our own. I think we were bright students, but we weren't brilliant and we were only juniors in college."

"The Professor was extremely disorganized. Worst of all she was not at all clear in her directions. I shared the responsibility with another student and we were constantly being told to redo what we had already done because it wasn't exactly what she wanted. Unfortunately, I don't think she knew what she wanted when she would give us the assignments . . . She yelled at us and blatantly ignored us."

"The professor in charge was to provide me with material from her regular class time, which I was to teach intensively in drill sessions. This worked all right for a while, but the situation rapidly became abusive. She never provided me with any material, so I had the option of not teaching the class for the day (or several days), which I felt was robbing the students."

"The alumnus wanted to sponsor an internship in his company, but ultimately he didn't care if I had work to do or not. It was a very frustrating experience and a waste of my time."

There was no institution or type of program that experienced these negative outcomes routinely. They were not the function of program or institution but resulted from a lack of attention and caring by individuals and an occasional bad match.

What becomes very clear is that it makes a difference with whom someone works, rather than what they do, where they do it, or when they do it. Further analysis in this chapter will look at the number of experiences, when in the academic year they take place, whether the experience happens on campus or off campus, etc., none of which correlates to the benefits of these programs in comparison to the human interaction. Clearly when an adult takes on a sense of responsibility and ownership for the experience and mentors the student, the value and the benefit are acknowledged as being important. The importance exists whether it eventually is one of many opportunities that blend into the fabric of an undergraduate experience or whether, as many cited, it is a once-in-a-lifetime happening, the meaning of which cut across all aspects of one's personal/professional development.

Next, a general review of the type of experience was conducted to see which experiences link up with which outcomes. (See Table 4-2.) More than two-thirds of the respondents indicated that their experience was related to their major. Virtually all of the respondents from the Dana institutions and the University of Rochester agreed or strongly agreed that their experiences were related to their major in contrast to participants in The Cornell Tradition program. It was noted earlier that because of Tradition fellowship requirement that students prove they've worked a significant number of hours during the academic year, many Tradition students could only fulfill those expectations working in routine positions such as the dining halls.[1] The results of this marked difference in experience between Cornell and most of the other programs, is that Tradition participants were less likely to say that their experience affirmed their choice of major or career plans or influenced their decision to attend graduate school or helped them gain professional contact, etc. On the other hand, all students noted with frequency that placements resulted in the acquiring of specific skills in such areas as problem-solving, independence, communication, team work, etc.

> "I participated in research with a faculty member. Mostly, I translated from a foreign language to English. Then I had to figure out the math explained in the text. This experience encouraged me to pursue a graduate degree. I now have an M.S. I don't think I'll go back for a Ph.D.
>
> I've taught at a junior college for the past nine months. Currently, I am the(volunteer) coordinator of volunteers for the city. I also teach water aerobics three times a week. Do I use much of my foreign language skills? No, but I do use problem solving skills I learned from my Dana experience.
>
> In six months all of this will change as my husband and I are moving to London."
>
> "I used my Dana grant to do community service in a small village in Senegal. This experience led me to pursue jobs in the public health sector following my graduation. I worked on a funded project for one year. Because I could advance professionally at that project without further field experience or a master's, I chose to temporarily pursue another interest. I hope to return to public sector work for developing countries

---

[1]The original Tradition program called for 360 hours of work in the academic year for continuing students. This was reduced to 300 in 1985 and eventually 250 in 1987. Hours for new students were eventually brought down to 200.

Table 4-2: Type of Experience

Experience	%	Significant Differences
Related to Major	68%	Cornell less likely than Dana and Rochester
Faculty were involved	44%	Dana more likely than all others
Alumnus(a) involved	11%	
Worked 11-15 hours/week during the academic year	50%	Cornell-56%, Dana-45%, Duke-N/A, Rochester-56%
Worked greater than 15 hours/week	18%	Cornell-28%, Dana-8%, Duke-N/A, Rochester-16%

following graduate school. I would not have these occupational goals without the Senegalese experience that my Dana grant made possible.

Even though I am not doing that type of work at this time, I use the attributes I learned and reaffirmed while in Senegal daily, specifically my ability to communicate, adapt to different situations, and be more self confident and self reliant.

I look forward to seeing the results of this survey. I am grateful that grants such as the Dana grant exist. My experience further enhanced my undergraduate learning experience."

As mentioned before, a large number of students, particularly at Dana schools, had an experience which involved a faculty member (44 percent). A much smaller group (slightly more than 10 percent or 300 students) had an experience with an alumnus. Although a faculty member was likely to influence the way a student thought about and related to the Academy, particularly as it pertained to graduate education, the value of the experience was not based on the status of the supervisor as faculty member but rather on the interest, care, attention, and support that the person gave. Fully 69 percent of those who said they would or were attending graduate school also indicated that they were influenced by a faculty member.

"The Dana internship program offered me a chance to work one on one with a faculty member outside my major. It proved to be an extremely important relationship during my time at Bryn Mawr. The faculty member with whom I worked was a mentor, friend, and colleague. He helped me decide on graduate studies and facilitated my choices. He offered me new horizons and was unfailingly supportive. His faith in my abilities has often exceeded my own. After graduation, he continued to keep in touch with me and asked me to cover classes while he was at a conference. He was an unofficial reader of my master's thesis and was the person who suggested that it could be published. I now have secured a book contract."

"The professor that I worked with was wonderful. She instilled in me a love of learning and research. She made me believe that I had what it took to go to graduate school and succeed if that is what I desired. My experience in the program helped me grow as an individual and to have faith in my abilities. I eventually hope to go to graduate school full-time."

Half the students worked between 11 and 15 hours a week during the academic year and there seemed to be something magic about more than 10 hours but no more than 15 not only as it pertained to the frequency of the experiences but also to the way in which participants responded. Again, the Cornell data demonstrated demand for a large number of hours during the academic year. More than one out of every four Cornell respondents indicated that s/he worked more than 15 hours a week. That demand was viewed by some as being a detriment to their academic work, to their personal development, and cast a pall upon the overall experience.

> "I needed both a summer fellowship and an academic year fellowship. These monetary awards made my career possible at Cornell. Without such assistance, my loans would have been very high. By reducing my debt while studying at Cornell, I was able to continue on to graduate school. The only drawback to the academic year fellowship was the amount of time I had to spend studying and working. Because of an extremely difficult semester, my grades suffered due to the additional work obligations for the fellowship program.
>
> Overall, I support the Fellowship Program and look forward to continuing to support it."
>
> "Being part of the Cornell Tradition was a terrific help for me, in addition to providing me with wonderful personal and professional experiences.
>
> However, I would like to note that not only did I work at my Tradition job during the academic year, I also had to take a second job to meet my needs. I ended up working 22 to 32 hours a week during the academic year, in addition to studying full-time at Cornell—extremely demanding. My grade point average definitely suffered, but worse, my health suffered even more. By the end of my senior year I had a severe case of bronchitis and was physically so run-down that I broke two ribs coughing. After that I developed chronic insomnia. The negative effects on my health are still with me today. I mention this because I am concerned for current and future students who require financial aid to attend our colleges (I was on full financial aid). I know the costs of an education are even much higher than when I attended Cornell. Although programs like the Cornell Tradition are definitely a big help, other solutions are needed. So, in answer to question #5, my GPA did not suffer as a result of participating in the Cornell Tradition, because I would have had to work anyway."

A thread running throughout the anecdotal comments of the participants was the degree to which these experiences "taught one how to behave, took the mystery out of what was expected of someone in the work place, provided a reference point and a source of self confidence from which one could then emerge and then take on additional challenges."

> "The Dana internship program was invaluable to me and my career. After graduating from DePauw in 1989, it took me only four weeks to find a job. Potential employers were impressed with my three-month internship, and I am convinced that it got my foot in the door many places. My Dana internship was an integral part of my growth from student to professional.
>
> I hope someday to be in a financial position to sponsor a scholarship or internship like the Dana program. Thank you."

This was one of those benefits that again cuts across all programs, by type and nature, at all institutions. There was no particular correlation identified other than the opportunity to have a responsible position. On the one hand, it sounds simplistic to say that these experiences helped students understand what role they were to play as professionals in the workplace. On the other hand, if one understands that for many of these students it was their first ever, full-time, responsible post—for most of these students it was their first professional position—it begins to make sense why this outcome was both very real and very important.

Some institutions, such as Bates, Bucknell, and Davidson, worked hard to provide a very comprehensive set of multiple experiences in different placements, different settings, on and off campus during the summer and academic year, set in a foreign culture, etc. Others, such as Ithaca College, more frequently limited the benefit of these experiences to a one-time event. In analyzing the results, it is clearly very important to ascertain whether focusing significant amounts of time, money, and attention on a small number of individuals with multiple opportunities produces a significantly different outcome than limiting a participant to one such career-related, educationally purposeful experience during their undergraduate career. While of course those who had multiple placements had more memories, more exposure, greater chance to be influenced, there are no data that verify that there are significant developmental differences in the value to the individual if they have multiple experiences versus only one. Therefore, if funding opportunities and placements are limited, as they most certainly will be at most institutions, it would seem to be efficient and of greater value to spread those experiences over the largest number of students possible rather than focus the attention on a small number of individuals. As highlighted in Chapter 1, only the impact on the decision to enroll, closeness to alma mater, and debt reduction were cited as significant advantages to having more than one of these educationally purposeful, career-related experiences. One way to accomplish this is to let the market dictate the product—that is, create a series of entry requirements, similar to The Cornell Tradition, and let the market dictate by student demand in how many experiences one participates. There just is no convincing evidence that an institution gets the "bang for the buck" by mounting experience upon experience for an individual undergraduate.

Finally, in looking at different types of experiences, the data was segmented by those who had a summer experience only, in contrast to those who had an academic year experience only, in contrast to those who had both. Table 4-3 shows this distribution. Most of the outcomes were predictable. Most of the on-campus experiences happened during the academic year although there were a number of summer research internships. The public service and for-profit experiences for the most part were summer although again some part-time academic work took place in social service agencies during the academic year. Careers were affirmed more frequently by those who had summer experiences, majors by those who had academic year experience. Because of the likelihood of the off-campus placement being part of summer experiences, outcomes like the development of a sense of civic responsibility followed suit.

Because there were only three programs that provided summer-only experiences—Duke, Wesleyan, and Davidson, two of which were off campus only (Duke and Wesleyan)—some interesting outcomes were observed. The sample here is of course very small and conclusions need to be drawn with caution. Wesleyan and Davidson participants were more likely to be transfer students; hold a public service/not-for-profit position; state that the experience contributed to their sense of civic responsibility; indicate that the experience enabled them to improve their self confidence and become more independent; indicate that the internship enabled them to have an experience they would not have had if it weren't paid; state that their ability to do volunteer work while in college was limited by their need to earn money; agree that participation in the program made them feel closer to the alma mater; and, finally, affirm that the decision to

Table 4-3. Time of Program Experience

Academic Year	Summer only	Both
Canisius**	Davidson	Bates
	Duke	Birmingham-Southern
	Wesleyan	Bryn Mawr
		Bucknell
		Cornell
		DePauw
		Dickinson
		Franklin & Marshall
		Furman
		Ithaca
		Parsons/New School
		Rochester
		Smith
		Vassar
		Union
		Wheaton

**Any finding here might not be because of the academic-year-only nature of the program, but because there is only one school in this category.

attend Wesleyan and Davidson was influenced at least in part by the opportunity to participate in this program, in contrast to those in the Duke Futures program. However, Duke Futures respondents were significantly more likely to have had an experience in a for-profit organization; state that that experience contributed to the development of their technical skills; and disagree that participation helped them gain professional contacts (although two-thirds of both groups agreed). One can only speculate why Duke Futures graduates were more likely to disagree than Wesleyan and Davidson graduates regarding gaining professional contacts. It may have to do with expectations that were created by the program. The Duke Futures program was sold to participants, or at least created the expectation, that professional contacts would be gained, and even though they were in most cases, when they weren't it caught the attention (and the ire) of program participant. Finally, Duke Futures participants were more likely to say that their experience involved working with an alumnus(a) than those from Wesleyan or Davidson. There was no difference between summer experiences on or off campus.

Among those institutions that offered programs both during the academic year and summer—excluding Cornell which is described in detail in a separate section of this chapter—students who had experiences at the Dana schools were more likely to be involved with a faculty member on campus, with all of the benefits that link with that experience, such as work related to major, gaining a sense of independence, and having an opportunity that they would not have been able to have if it was unpaid. The Rochester students were more likely to be involved in not-for-profit organizations with the attendant outcomes described earlier, such as being more likely to develop a sense of civic responsibility as well as a sense of teamwork and enhanced communications skills.

In sum, therefore, all these experiences helped students prepare for the real world. People count more than placements or duties assigned. More than 15 hours of work a week during the academic year can be a deterrent to a positive experience as well.

Turning to developmental outcomes, both academic and career as well as personal, the data are rich. (See Table 4-4—Developmental Outcomes, Results, Etc. of Program Participation.)

Only 11 percent of the participants said that the experience changed the focus of their major while more than a third said that participation affirmed their choice of major. It is likely that these outcomes, along with work that was related to the major, are limited to those in the upperclass years, particularly juniors and seniors, because as was noted earlier two-thirds of the participants in all of these programs were juniors and seniors, students who were likely to have chosen majors. In that light, these response rates make sense. It would be unusual for a large number of upperclass students to change the focus of their majors when the program was limited to upperclass students. Once again, because of the nature of The Cornell Tradition experience (heavily administrative internships in places like offices and dining halls), alumni from Rochester and Dana schools differed from Cornell in response to this question.

A number approaching half (well over 1000 students) indicated that these experiences affirmed their career plans. Again, Cornell participants were significantly less likely than those from other institutions to cite this outcome due to the makeup of many Cornell placements. A quarter of the alumni said that the experience influenced their decision to attend graduate school (strongly linked to faculty influence). Half of all respondents said that they had planned or in fact

Table 4-4. Developmental Outcomes, Results, Etc. of Program Participation[a]

	All	Cornell	Dana	Duke	Rochester
Program changed the focus of their major.	11%	8%	12%	13%	18%
Affirmed choice of major (Rochester and Dana respondents more likely to indicate this than Cornell and Duke participants).	37%	20%	48%	33%	49%
Affirmed career plans (Cornell participants significantly less likely than those from other institutions to cite this outcome).	43%	28%	46%	47%	43%
Influenced decision to attend graduate school. (Cornell respondents less likely to cite this as an outcome than all other participants.)	25%	16%	31%	27%	31%
Planned to or had attended graduate school.	62%	55%	68%	50%	68%
Helped gain professional contacts. (Rochester, Duke, Birmingham-Southern, Ithaca, Canisius, DePauw, Wesleyan, and Smith participants cited this outcome more than Cornell participants.)	45%	32%	49%	63%	58%
GPA increased as a result of participating. (Dana recipients more likely to cite this outcome than Cornell and Duke participants.)	18%	18%	20%	7%	15%
Internship enabled them to have an experience they could not have had if unpaid. (Cornell respondents less likely to cite this benefit than other participants. Duke respondents less likely than Dana participants.)	71%	55%	82%	72%	70%
Ability to do volunteer work during college was limited by the need to make money. (Duke respondents were less likely to cite this obstacle than respondents from Cornell or Dana schools.)	70%	74%	72%	54%	56%
Contributed to sense of civic responsibility (Cornell and Rochester participants more likely to cite this outcome than all others.)	30%	37%	21%	12%	30%
Experience improved self confidence.	50%	41%	54%	49%	54%
Experience attributed to the development of adaptability/flexibility.	40%	46%	37%	36%	35%

cont'd

Table 4-4. Developmental Outcomes, Results, Etc. of Program Participation[a]

	All	Cornell	Dana	Duke	Rochester
Teamwork was an acquired skill.	26%	29%	24%	24%	35%
Developed communications skills.	46%	42%	47%	42%	61%
Acquired problem solving skills. (Participants from DePauw, Bucknell, and Bates cited this outcome more than all other respondents.)	36%	25%	44%	40%	37%
Developed greater sense of independence. (Dana respondents were more likely to cite this outcome than Reach participants.)	46%	45%	49%	44%	35%
Developed technical skills. (Cornell respondents less likely to cite this outcome than all other participants.)	35%	25%	41%	43%	39%
Indebtedness had been reduced. (Cornell respondents more likely to cite this outcome than all other program participants save those from Bates and Canisius. Lot of variation in expense as at other institutions.)	45%	82%	28%	17%	20%
Reported annual salaires. (Duke participants reported higher annual salaires than most other Dana schools. Cornell salaries were second highest.) See Table 4-5, page 113.	≥50k ≤20k	20% 11%	6% 24%	20% 3%	5% 19%
Influenced decision to enroll. (Cornell, Bucknell, and Canisius more likely than all other participants to cite this outcome.)	10%	19%	6%	19%	5%
Participation made alumni feel closer to alma mater. (Cornell participants more likely to cite this outcome than Duke, Ithaca, and Rochester participants.)	48%	65%	40%	26%	34%
Involved in alumni club in local area. (Duke alumni were more likley to cite this outcome than participants from all other institutions.)	30%	36%	24%	42%	20%
Received awards or scholarships as a result of participation. (Cornell respondents more likely to cite this benefit than other participants.)	18%	28%	14%	6%	12%

[a] Respondents citing this outcome agreed or strongly agreed.
[b] The only Cornell alumni used for this comparision were those who participated in the program in or after 1987.

were attending graduate school. Again, Cornell respondents were less likely to cite this outcome than participants from other institutions. Almost half of those who responded to the alumni survey said that their experience helped them gain professional contacts. Mix of schools for which this was the case is curious. It includes Rochester, Duke, Birmingham-Southern, Canisius, DePauw, Wesleyan, and Smith; once again Cornell being the outlier. There is no particular programmatic thread that runs through the experiences and placements at those institutions that would logically link these schools and this benefit. The Duke experience was heavily off campus during the summer with the assistance of alumni. Canisius was entirely on campus. Wesleyan was primarily in the summer but Rochester, DePauw, Smith, Birmingham-Southern, and Ithaca were both on and off campus, during the summer and academic year, some with multiple experiences, others only a one-time event (Ithaca), so there doesn't seem to be a particular type of program that produces this outcome. Alumni were most appreciative of this benefit when it occurred.

> "In the entertainment industry it is truly about who you know. I knew many people before I even graduated because of my internships. At 25 I am one of the youngest casting directors in NYC and I owe a large part of that success to my internships. I figured out exactly what I wanted through these experiences so I knew what to pursue after college. I am the perfect example of why programs like the Dana scholarship are so important. Now I take it upon myself to continually train new interns."

There is no substantial evidence that these experiences have a positive value on one's academic performance. As noted in Chapter 2, this is consistent with the rest of the literature on this subject. Less than 20 percent of the students cited this outcome; however, when it was cited it was positive (as opposed to having a negative impact on the GPA) and the Dana recipients were more likely to cite this outcome than respondents from Cornell or Duke. The common programmatic element producing this result at the Dana schools would point toward the relationship of the faculty member and the fact that most experiences were on campus and academically related. Finally, as one would anticipate, Cornell respondents were less likely to note that the Tradition placement enabled them to have an experience they could not have had if it was unpaid than Dana or Rochester respondents. Duke respondents were less likely to cite that their ability to do volunteer work was limited to the need to make money than respondents from the other institutions. The former, again, is a statement about the nature of a large number of the Cornell placements. The latter has to do with the academic profile of the Duke student—that is, the Duke student body tends to be more affluent than most if not all of the other institutions' enrollees.

On the personal development front, there were two major findings—one noted earlier, when the experience contributed to the development of a sense of civic responsibility, and the second being the development of self confidence. On the first account, Cornell participants were more likely to cite this as an outcome than all others. This follows closely from the founding principles and goals of The Cornell Tradition and the extent to which the Cornell program, particularly since 1988, had striven to build community among Tradition participants and sense of obligation and commitment to the Tradition, to the alma mater, and to future generations of students. The reason why the civic responsibility outcome is important can best be described by understanding what other outcomes link significantly to this benefit. Other significantly positive experiences and outcomes include feeling close to the alma mater, gaining professional contacts, acquiring the skills of adaptability and flexibility, improving self confidence, gaining a sense of value of teamwork, improving communications skills and technical skills, and acquiring a sense of independence. These are most likely experienced when involved with an alumnus or alumna in a public service, not-for-profit setting. And, finally, this outcome links to early participation in one's undergraduate career and to acknowledging that opportunity to participate in the experience and influence the decision to enroll.

The improvement of one's self confidence is the equivalent acquired attribute most frequently cited in the anecdotal responses provided by the alumni regardless of institution and regardless of the number of times one participated in the program. Self confidence turns out to be an important outcome variable from an employer's perspective. Research conducted at the University of Rochester in the summer of 1994 found employers citing that hands-on, practical experience that led to an improved and enhanced level of self confidence was an important and sought-after attribute when permanent hiring decisions were being considered. Half of the respondents to the alumni survey indicated that improved self confidence was one of the most important benefits of participation. Greater than half of the Dana and

Rochester participants noted that outcome. Here the analysis of common experiences is interesting. The most important criteria has to do with when one participated in the program. For instance, those who had a freshman experience were least likely to develop the skill of self confidence while over half of those who participated, at least as sophomores and had a public service, not-for-profit experience, indicated that self confidence was one of the three most important skills developed. Females more than males who participated in their junior and senior year developed self confidence. However, the chances of developing self confidence for males increased if they had a teaching assistantship with a faculty member or a for-profit experience.

> "Cultural awareness and appreciation without question would be number one. Then perhaps adaptability and a lot of creativity thrown in and then independence. My self-confidence in terms of communication skills and socializing was in a way brought into question - I realized I was too caught up in trying to present myself as someone I wasn't."
>
> "I ended up pursuing a career in marketing with a great deal of confidence. Confidence I would not have had without the exposure to the 'business culture' that I received through the internship. You are so young and inexperienced in college that it is almost like a guessing game to choose a career. I was very uncertain about which path to take, and my Dana internship gave me the clarity and confidence in the decision making process."
>
> "The experience increased my independence, self-confidence and understanding of how to behave in the workplace. It was my first experience in the workplace and I thoroughly enjoyed it. It allowed me the opportunity to live in Boston while participating in two unique communication job experiences. It sparked my interest in the non-profit sector and made me focus on what I was looking for in a career."

Other enhanced skills or personal development outcomes as noted in Table 4-4 include: 40 percent of the respondents cited increased adaptability and flexibility as a result of the experience; 26 percent acquired the skill of teamwork;

> "This experience [at a public service legal organization] was the first time that I had to truly depend on others to make something work out smoothly. There were three of us who were working on cases, depending on each other to do different aspects of one case. Good experience for someone who likes to do everything on her own. Control freak."

Forty-six percent enhanced their communications skills; 36 percent acquired problem solving skills.

> " . . . the success of this program was greatly dependent on the person for whom I worked. He continually demanded improvement and a high volume of work. I learned how to get information from bureaucratic corporations, solve problems, and work independently."

Forty-six percent developed a greater sense of independence; and 35 percent developed technical skills. While some of these outcomes match up with some specific programmatic experiences, most in fact don't. Problem solving is most frequently mentioned in institutions such as DePauw, Bates, and Bucknell because of the correlation of having a summer research, science-oriented internship and citing the improvement of problem-solving skills. Students who had a research assistantship with a faculty member, in contrast to those with other experiences on or off campus, cited a greater sense of independence as one of the benefits of participating in that opportunity. For example, Dana respondents were more likely to cite this outcome than participants in the Reach for Rochester program where research assistantships with faculty members were infrequent at best.

The final individual benefit examined was financial, notably, debt reduction and annual salaries. In regard to the former, debt reduction was the most frequently cited institutional goal as defined in the institutional survey. It also was for the Dana schools the most frequently cited goal in their proposals to the Foundation. It turned out to be a very difficult outcome to achieve. Slightly less than half of all program participants (45 percent) indicated that the experience had helped them reduce the amount of money they would have had to borrow otherwise in order to complete their undergraduate education. However, this aggregate figure is misleading, for over four-fifths of the Cornell students cite this as an outcome because Cornell makes up, as noted earlier, a third of all the program respondents. With Cornell excluded, just over 25 percent of all other respondents cite this as a benefit for participating in these programs. In fact, Cornell respondents are statistically different from others save those from Bates and Canisius.

> "The Tradition allowed me to accept a position upon graduation that I could not have 'afforded' had I had all the student loan I 'should have'

had. Additionally, it gave me an 'incentive' to work, as my efforts to pay for my own education were recognized. It is also the only fund that I would contribute money to (at Cornell), as I feel it recognizes need, willingness and efforts towards helping yourself and others."

"I believe that the Cornell Tradition was a valuable experience, but I must say that I would have worked anyway. An immediate and lasting effect of the experience was the pride I felt in being selected to participate. The main help from the program was the reduction of debt. It allowed me to seriously consider graduate school, too. Although I did receive (later) funding for grad school, I at first did not have any support, and would not have wanted to acquire any more debt, if I did, indeed, have several large outstanding loans."

On the second financial indicator, one-way comparisons showed that Duke program participants reported significantly higher annual salaries (greater than $40,000) than alumni respondents at Smith, Wheaton, Canisius, Vassar, Wesleyan, DePauw, Birmingham-Southern, and Bryn Mawr. The Cornell Tradition participants reported a significantly higher annual salary (just under $30,000) than Smith, Wheaton, Canisius, Vassar, and Bryn Mawr graduates.[2] Bates and Davidson salaries were significantly lower than Cornell's but low response rates meant that the numbers were not statistically significant. (See Table 4-5.)

There were two distinct institutional benefits examined—the first being recruitment, the second having to do with alumni relations. Recruitment, that is, influencing the decision to enroll based on the opportunity to participate in these programs, was one of the institutional goals frequently boasted in the proposals to the Dana Foundation. It, like debt reduction, however, turned out to be a very difficult objective to accomplish. In fact, overall only 10 percent of the respondents indicated that the opportunity to participate made a difference in where they became undergraduates. Those respondents can almost exclusively be tracked to only three institutions: Cornell, Canisius, and Bucknell. The common denominator here is that those institutions had a significant percentage of freshmen participating; therefore students didn't have to wait for a year or two or even three before becoming actively involved and the schools promoted that participation heavily in their recruitment literature. Engagement in the experiential opportunity or benefit began either in the summer after the

---

[2]The only Cornell alumni used for this comparison were those who participated in The Cornell Tradition program in 1987 or after.

Table 4-5. Annual Salary Table by Program

Annual Salary	Cornell	Duke	Rochester	All Others
Less than $20,000	64 10.6%	3 3.1%	11 18.6%	187 24.0%
$20,000-30,000	160 26.4%	21 21.4%	23 39.0%	320 41.1%
$30,000-40,000	177 29.2%	33 33.7%	15 25.4%	159 20.4%
$40,000-50,000	86 14.2%	21 21.4%	7 11.9%	70 9.0%
$50,000-70,000	75 12.4%	14 14.3%	3 5.1%	29 3.7%
$70,000+	44 7.3%	6 6.1%		14 1.8%
Total	606 39.3%	98 6.4%	59 3.8%	779 50.5%

senior year of high school, before formally matriculating as a freshman, or at some point during the freshman year experience.

> "The Dana program was a marvelous opportunity for me. It allowed me to explore a field removed from my original major, influenced me in changing my major, and inspired my later work in that field. Additionally, the program was a powerful inducement in my decision to attend Bucknell. The fact that I was chosen for an offer made me feel the school was pursuing me in an active manner. Overall a four star program."

Other schools talked about putting these experiences in their admissions literature or in their videos (Smith) but also admitted in the institutional survey that they were not able to demonstrate or document positive, observable enrollment outcomes.

The alumni relations benefits to the institution centered on two outcomes: one, whether the experience made participants feel closer to their alma mater and secondly whether participants were involved in local alumni club activities after graduation. Forty-eight percent of all respondents indicated that they felt closer to the alma mater as a result of their experience. Cornell's participants were more likely to cite this outcome than respondents from Duke, Ithaca, and Rochester. There were two programmatic activities that correlated with this outcome. One was students who had an early experience, that is, became engaged in the program and opportunities in their freshman or sophomore years. And, second, programs that consciously and deliberately worked at developing a sense of community (Cornell, Bucknell, Dickinson, Furman, and Wheaton). One of the benefactors of The Cornell Tradition program said during an interview about why he financially supported the Tradition that he "sees the Tradition as an elite corps of alumni, elite because they do good things, working in community service, etc." In fact this alumnus feels that the goal of the Tradition should be to have everybody at Cornell be in the Tradition program. He sees a real cause and effect at work. The Tradition is creating community; community is dedicated to service; service is a membership card to belong to the community. He would like to make the Tradition all-inclusive because he sees work as being an honorable thing and community service as being an honorable thing and when combined they meet his definition of a prestigious institution, one to which people are connected for the rest of their lives.

The second alumni relations benefit had to do with involvement in local alumni clubs after graduation. Almost a third of the respondents indicated that they were involved. This likely represents two to three times the number of young alumni from those (or any institution) that would cite involvement within five years of graduation. Duke alumni were more likely to cite this outcome than participants from all other institutions, which is curious given some of the other responses from the Duke Futures participants such as whether the opportunity allowed them to gain professional contact, reduce indebtedness, etc. One can only speculate that this outcome is more linked to the institutional difference of Duke versus the other colleges and universities in this study than it is to participation in the Duke Futures program.

The research revealed the following profile of characteristics and program attributes for those who responded that they were now involved with their local alumni club.

- There was no gender difference or difference between freshmen or transfers in whether program participants were presently involved.
- Those who were research assistants with faculty were significantly less likely to be involved in a club. The contrary was true for those who had administrative internships on campus or an

experience with a not-for-profit organization (probably a Cornell-driven outcome).

- If the respondent noted the experience affirmed his/her career plans or contributed to a sense of civic responsibility, he/she was significantly more likely to belong to an alumni club.
- Finally, participants who said they acquired the skill of adaptability and flexibility were also significantly more likely to currently belong to an alumni club.

II. Cornell Research Results

Cornell Tradition participants, as noted earlier, represented more than a third of all respondents. For this reason there are a number of observations about The Cornell Tradition that can validly be made that are not possible with the other programs. First, a large number of responses allows for a detailed profile of the Tradition participant. As is the case with involvement in these types of experiences and with responding to the surveys, women dominate the Cornell responses. Women were more likely to participate in an internship in a public service, non-profit organization; say that the experience changed the focus of their major (although it didn't change the focus at all for greater than 90 percent of all participants); say that their experiences affirmed their career and helped their self confidence; and agreed that they gained contacts and that their ability to do volunteer work was limited by the need to earn money.

Transfer students who participated in The Cornell Tradition were more likely than those who began their Cornell career as freshmen to agree that participation in the Tradition helped them to gain professional contacts; have an experience related to their major; and state that volunteer work during college was limited by their need to earn money and that the opportunity to participate impacted their decision to attend graduate school. Because transfer students are most likely upperclass, these issues which tend to arise later in an academic career, such as major and decisions regarding graduate school, made sense.

Non-traditional students, comprising almost 5 percent of The Cornell Tradition participants, are defined here as students over 25 years of age upon entrance to Cornell. They were, as a group, more likely to agree that they were limited in their ability to do volunteer work because of a need to earn money and stated that the involvement in the Tradition affirmed their major more frequently than traditional age students. There is a large overlap between non-traditional students and transfer students (nine out of ten). Fully 14 percent of all transfer students were non-traditional.

The final profile characteristic analyzed was ethnicity of the respondents with particular attention paid to those racial ethnic groups that have been traditionally underrepresented in the Cornell student body—African American, Hispanic American, and Native American. In general the underrepresented minority students in The Cornell Tradition (greater than 10 percent of all Tradition participants) did not cite significant differences in their experience from those of other participants, except that underrepresented minority students were more likely to say participation in the Tradition significantly influenced their decision to attend graduate school, led to a sense of civic responsibility, and improved their self confidence more often than Caucasian or Asian American respondents.

One of the benefactors of The Cornell Tradition, when interviewed regarding his own philanthropic support for the Tradition, said "The Tradition is like apple pie, work and community service." To him the Tradition is an eclectic set of experiences. It is composed of opportunities as distant as the rural mountains of Colorado all the way to the cosmopolitan streets of New York. The gamut of experience which attracts a diversity of people, cultures, ideas, and values is something he sees as being very typical of the Cornell undergraduate and very unlike the campus atmosphere found at most other competitive, high-cost institutions. The Tradition, because of its makeup of both participants and experiences, lets Cornell talk about itself. The Tradition, he says, is Cornell.

Minority student participation was a major programmatic goal of the Tradition. In fact, another Cornellian who financially supports the Tradition when interviewed boasted that her proudest contribution as an alumna was her connection with the Cornell Black Alumni Association (CBAA), a 4000 strong organization with 10 chapters around the country. She was proud of being able to help the CBAA raise $60,000 for two named fellowships. In addition to the Tradition Fellowships, the CBAA has established a scholarship endowment fund beginning with over 100 alumni pledges and payments in excess of $150,000. The CBAA's scholarship endowment fund newsletter (*Sharing the Vision of Our Education*) called the CBAA Cornell Tradition Fellowship a bridge to the future.

As highlighted in Chapter 3, the Tradition program changed focus and direction after a review in 1987. For the purpose of this study, then, Tradition participants who graduated in 1987 and before were segmented from those students who were participants in the program exclusively starting in 1988 and thereafter. Any student whose experience bridged both was excluded from this analysis.

The profile of the student in the original Tradition program was slightly different. There were more transfer students participating due to the desire to build up the Tradition program's numbers in the early years. Placements were also somewhat different after 1987. There was more experience in the public service, not-for-profit sector, and with faculty than in the earlier years. The result of those placements was that the new experiences were most likely to change the focus of major, affirm career, and contribute to a sense of civic responsibility. (Twice as many reported this last effect when participating exclusively in 1988 and thereafter than those who participated only in 1987 or before.) This is a particularly important outcome given that the Cornell Tradition desired to develop a strong sense of citizenship. Further, participants in the modified program cited that they learned adaptability/flexibility, teamwork, and communication skills more often than those in the original program. The recruitment value of the Tradition has always been held as a strong program component. Changes that were made after 1987 in fact produced twice as many participants who indicated that the opportunity to participate influenced their decision to attend Cornell than those in the first five years of the program. These later participants also said that the faculty influenced their plans after graduation (consistent with the increased involvement with faculty). Finally more of the participants in the new program said that debt reduction was a frequent outcome (87 percent versus 79 percent). On the other hand, participants in the original program were significantly more likely to agree that their ability to do volunteer work during college was severely limited by the need to earn money and were more likely than those who came later to say they did not have plans to go on to graduate school.

Table 4-6 documents the cause and effect of the Tradition programmatic components and the experience attributes and personal development outcomes of the participants. All of the observations cited are statistically significant. Clearly, as noted earlier, when the experience contributes to a sense of civic responsibility, a lot of good things happen. The obvious linkage of that positive outcome with a public service, not-for-profit placement, is logical. There is a strong statistical

Table 4-6: Cornell Experience, Attributes, and Personal Development Outcomes. (All Statistically Significant)

	Worked in governmnet agency	Worked for faculty member	Faculty influenced plans following graduation	Work was related to major	Experience involved alumnus(a)	Experience influenced plans to attend graduate school	Opportunity influenced decision to enroll	Changed locus/affirmed major	Affirmed career plans	Contributed to sense of civic responsibility	Reduced indebtedness	Helped gain professional contacts	Felt closer to alma mater	Able to have experience could not have had if not paid	Acquired skill of adaptability/flexibility	Improved self confidence	Acquired skill of teamwork	Improved communication skills	Acquired skill of independence	Now involved with local alumni chapter	Totals
Research assistantship with faculty																		X		X	2
Internship in public service	X		X	X				X	X	X	X	X	X								9
Internship in a government agency	X			X					X				X	X							5
Internship in a for-profit organization					X			X							X						3
Contributed conference papers/ attended conferences													X								1
Experience contributed to sense of civic responsibility					X	X	X					X	X	X	X	X	X			X	10
Experience influenced plans to attend graduate school		X	X	X	X	X						X		X						X	8
Experience involved faculty		X		X								X		X						X	5
Experience involved an alumnus(a)					X							X	X		X				X		5
Experience helped gain professional contacts					X		X					X	X	X		X	X	X			8
Work was related to major				X									X	X			X	X			5
Felt closer to alma mater as a result of experience							X						X		X	X		X		X	6
TOTALS	2	2	2	5	5	2	3	2	2	1	1	6	8	6	4	3	3	4	1	5	

correlation also between having an experience which helps gain professional contacts and other positive outcomes. Self confidence matches up with feeling closer to the alma mater; having one's work relate to the major; gaining professional contacts; and, the big winner, of contributing to the sense of civic responsibility.

III. The Dana, Duke, and Rochester Experience

Because of the dominance of Cornell in some of the areas of these experiences, responses, etc., a separate analysis, not including Cornell, was conducted to test three distinct areas of these experiences. These conclusions are based on a sample of about 1750. The segmentations include time of participation, whether the experience involved an alumnus(a), and finally whether the experience was on or off campus. As Table 4-7 notes, those who had a freshman year experience, whether or not they participated as upperclassmen, were less likely to be placed in a public service, not-for-profit agency, with a for-profit organization, or involved in work related to their major. On the other hand, because of the early exposure in their undergraduate career, their experience did change the focus of their major, influence their decision to enroll, and ultimately helped to reduce their indebtedness more so than those who participated in the upper-class years.

Working with an alumnus(a) as a sponsor, job developer, or supervisor involved internships in the public sector whether not-for-profit, including government agencies, as well as for-profit organ- izations. This makes sense because these opportunities are predominately off-campus placements. Working with or through an alumnus(a) meant that there was a significant difference between those placements linking to the development of a sense of civic responsibility versus others on campus with a faculty member. The alumni provided experiences also related to the enhancement of the skills of adaptability and flexibility and of self confidence, as well as providing opportunity to gain professional contacts.

Finally, regarding on- and off-campus experiences, on-campus were exclusively those at Bates, Dickinson, Canisius, and Bucknell, where outcomes that were cited for significant difference include influence in the decision to enroll and were probably driven by the fact that two of the four exclusive on-campus institutions were Canisius and Bucknell. (As pointed out earlier in this chapter, those are two of the three institutions that involved new freshmen in the experiential

Table 4-7: Dana, Duke, and Rochester Experience. Attributes and Personal Development Outcomes

	Less likely experience was part of service in not-for-profit	Less likely experience was in for-profit	Faculty influenced plans	Influenced decision to enroll	Involved faculty	Administrative intern on campus	Internship was with public service not-for-profit	Internship was with government agency	Internship was in for-profit organization	Affirmed major	Experience changed focus of major	Work related to major	Had positive influence on GPA	Received an award	Attend graduate school	Contributed to sense of civic responsibility	Acquired skills of adaptability/flexibility	Acquired self confidence	Gained professional contacts	Reduced indebtedness	Felt closer to alma mater	Totals
Freshman year participation	X	X	X	X	X	X					X									X		8
Alumni(a) involved in sponsoring, identifying, placing, or supervising							X	X	X							X	X	X	X			7
On-campus experience (Bates, Dickinson, Canisius, and Bucknell exclusively on-campus.				X						X			X	X	X					X		6
Off-campus experience (Duke and Wesleyan exclusively off-campus)												X	X	X				X	X		X	6
Totals	1	1	1	2	1	1	1	1	1	1	1	1	2	2	1	1	1	2	2	2	1	

[1] All results are statistically significant

opportunities and like Cornell realized a positive recruitment benefit.) Also, the benefits of exclusively on-campus placements included reduced indebtedness, affirmation of major, positive influence on GPA, receipt of an award, and plans to attend graduate school. By contrast, Duke and Wesleyan were exclusively off campus. Off-campus placements were more likely to be involved with work related to their major. These experiences produced a sense of closeness to the alma mater as well as gaining in professional contacts and improvement in self confidence.

Many of these outcomes match up consistently with the programmatic detail, outcomes, and attributes of The Cornell Tradition. This segmentation confirmed that the Tradition influence was not separate and distinct from those outcomes experienced at the other institutions. In other words, although it is true that Cornell dominates the responses, many of the outcomes are universal and not Cornell-specific.

IV. Women Participants: Single Sex vs. Co-ed Colleges

As can be seen from Table 4-8, there were some differences in the experiences of women participating in these educationally purposeful, career-related work opportunities based on a single sex versus co-ed environment. Notably, co-ed placements were more likely to be with teaching assistants or for-profit organizations versus placements at single sex institutions, which were more likely to take place in a public service, not-for-profit and government agency. Participants from Smith and Bryn Mawr were more likely to be transfer or foreign students. The former is likely an outcome of the Ada Comstock program, a significant program at Smith for non-traditional students, many of whom are transfers. The linkage between not-for-profit placement and civic responsibility is consistent with other outcomes observed as well as experiences involving an alumna. The other significant outcomes that participants from co-ed institutions reported, namely reduced indebtedness, contribution to a conference paper or attending a conference, and the acquisition of communications skills, do not logically link up with any programmatic distinctives of those Dana institutions.

Thus, the most interesting outcome of this analysis is not what was different but what was the same. That is, there was no significant difference between alumnae respondents from either a single sex or

Table 4-8: Women Participants at Single Sex Colleges (Dana Only) vs. Women Participants at Co-ed Institutions. Participants Profile, Program Attributes, and Personal Development

	Transfer student	Foreign student	Internship at public service not-for-profit agency	Internship with government agency	Internship as teaching assistant	Internship in for-profit organization	Contributed to sense of civic responsibility	Influenced decision to attend graduate school	Able to have experience would not have had if not paid	Experience involved alumnus(a)	Reduced indebtedness	Contributed to conference papers, attended conference	Acquired communication skills	Totals
Single sex colleges (Smith & Bryn Mawr)	X	X	X	X			X	X	X	X				8
Co-ed colleges					X	X					X	X	X	5

[a]Only significant differences noted

coeducational institutions on the development of such skills and attributes as adaptability, flexibility, self confidence, teamwork, problem solving, independence, and technical skills. It would appear therefore that the experience makes the difference, not the environment.

Certainly this doesn't by any means exhaust the outcomes, products, benefits, and results of the collective experience of more than 2700 participants who responded to the alumni survey. It does, however, begin to paint a very clear picture of the power of these experiences, the breadth and depth of the benefits, and the academic and personal domain in which these results reside. There are certainly measurable gains in personal growth and development. More than just occasionally, lives were shaped, changed, and made.

Chapter 5

Conclusions

As with any effort of this magnitude, there is a continuous number of choices to make and an ongoing tension between inclusivity and exclusivity. One often wonders and even worries whether the most significant finding has been left unattended. There are clearly much more data to mine and observations to make. The goal of this effort, however, was to look at what worked and what did not; what made a difference in people's lives and in the lives of the institutions sponsoring these opportunities. While we have commented about program structure and outcomes throughout Chapters 3 and 4, the loudest single voice of the participants underscored that these educationally purposeful, career-related, paid employment opportunities can and do make a tremendous difference in people's lives. They can be, not just occasionally, but routinely, career makers. These programs also can take on a life, shape a culture, and become part of the fabric of the sponsoring institution. That is certainly the case at Cornell University as well as at schools like Furman, Ithaca, and Wheaton.

The following then is a very brief review of what has been learned.

1. The enthusiasm, zeal, and commitment that program participants have to the program and to their alma mater can perhaps best be described by the tremendous response rate (greater than 50 percent) enjoyed in this research. Industry standards would have said that 25 percent or 30 percent would have been good and significant. Participants in these programs doubled that rate.

2. The goals of having an impact on accepted applicants' decisions to enroll at a particular institution and the opportunity to reduce indebtedness are very important institutional objectives. However, the experience demonstrates that these are two very difficult outcomes to produce.

Clearly, Cornell has been successful on both fronts. Only Canisius and Bucknell join Cornell in having a significant impact on the decision to enroll. It was obvious from the research that the promise of future opportunities in the upper-class years does not ring true enough to entering students to change their behavior. Rather, the program has to be specifically geared toward what the student can expect in his/her freshman year. This is an important outcome for other institutions to consider if in fact influencing the enrollment decision is a desirable and intended institutional outcome.

Likewise, regarding debt reduction, while many programs cited reducing indebtedness as an important goal, few were able to execute programs that have actually made a difference. Typically those that did either that touched a small number of students (e.g., Bates) or set their sights very specifically on debt reduction (e.g., Furman, Ithaca, Wesleyan, and DePauw joined with Cornell in reducing debt for a sizeable number of participants in the program.)

3. It would appear as though there is no particular difference between offering opportunities in the academic year or the summer or both; either on or off campus. (Time of participation in one's undergraduate career does seem to have some bearing on things like the development of self confidence and debt reduction.) However, one is struck by analysis after analysis and testimonial after testimonial of the influence that people have on the interns, the fellows, and the assistants whom they work with and for. It was not so much what they did, when they did it, or where they did it as with whom they did it. That is not to say routine work is as significant to one's growth and development as exciting career-related responsibilities. However, the following would appear to be true. An exciting, career-related educationally purposeful responsibility without a good mentor has less value and significance than a more routine or even more mundane set of responsibilities overseen and mentored by someone who cares.

4. The outcomes of developing a sense of civic responsibility, community building, and closeness to one's alma mater are definitely linked. As is noted in Chapter 4, the vast majority of those who said they developed a sense of civic responsibility as a result of their experience indicated that they had a public service, not-for-profit experience. When looking specifically at Cornell Tradition outcomes, one finds that those Tradition participants who cited that the experience contributed to their sense of civic responsibility noted numerous benefits including the acquiring of the skills of adaptability/flexibility, self confidence, teamwork, communications, independence, and technical knowledge, along with closeness to the alma mater. When

developing a sense of civic responsibility is an intended outcome of a program, many, many other positive experiences and benefits are gained.

5. The successful administration of the programs as well as the requirement to punctuate the work experience with a final report does play a role in participants maximizing the benefit received from their placement. In addition, of course, execution of the program impacts positively on such things as program continuation, fundraising, etc. Certainly in order for programs to continue there has to be an institutional commitment and understandably those who documented and monitored their outcomes were in the strongest position to argue for continuation and integration into the campus culture.

6. A number of these institutions enroll a very affluent student body. On more than one occasion it was noted that these educationally purposeful, career-related work opportunities which provided healthy stipends and/or significant hourly wages "balanced the playing field" for financial aid students. These needy students were able to take advantage of opportunities where otherwise the experiences would only have been available and accessible to more affluent populations. There was a very real sense of appreciation on the part of needy students at many, many institutions. They saw these programs had made their undergraduate educational experience whole.

As mentioned earlier in Chapter 1, the Academy today finds itself under siege in a variety of ways by many of its publics and constituencies. For example, Congressional hearings were held recently to examine the quality of undergraduate teaching at research universities; alumni are organizing nationally in response to what they view as bad decisions being made by their alma maters in regard to political correctness; taxpayer support for public higher education is waning; philanthropic support has turned toward elementary/secondary education; and, finally, employers are saying that graduates of our colleges and universities are not trained nor skilled enough to keep America's business and industry competitive in a global economy. While these career-related, educationally purposeful employment opportunities described and analyzed in this study are not put forth as a panacea, they do in fact respond to many of these concerns—concerns which fundamentally reflect an emerging lack of faith and trust in American higher education. These programs speak to quality interaction between faculty and students, to self help and work, to the development of citizenship and community, and to relevancy and employability.

These programs also create linkages with prior generations and bring attention and financial support to the Academy. One generous Cornellian was most eloquent in this regard.

> "With all of this going on, I did not have much money to give to Cornell; let's not kid ourselves, $10,000 a year is not much. Howsoever, I chose not to walk across campus, on a walker at age 90, and say 'shakily' to my children, 'here's a bench named after me!' The Tradition gave me the only chance to 'cast my bread upon the waters.' For all of the young people whom I have met, who have received fellowships in our family's name, I have said the same thing; 'it is hard for you to believe that you will ever have extra money in your jeans—but you will. And to the degree that you do, you have to promise me that you will help a student through school, or help a student reduce his/her loan obligation. My goal, now that I have taken care of my children, is to give the Tradition $100,000/yearly. My ultimate goal is to donate enough that the money invested will throw off enough income to provide a number of scholarships; i.e., a million dollars at minimum. God knows that all of us, deeply committed to higher education, are most grateful for those rare human beings who can donate $10, $15, $20, etc., a million dollars, but they want a name on the building; and my passion has become, 'well wait a minute—what we need is the funds to help kids through school.' As we concern ourselves with minority enrollment, these parents cannot look at a debt that will take them the rest of their lives to pay. The students can't get a job and then say 'Gee withholding, FICA, Medicare, and paying back my student loan, leaves me nothing.' So to repeat, Tradition kind of giving allows you to bring others into the giving process. I have always argued, that for every graduating student, we should ask that you donate (signed on a piece of paper) $5 the first year you are out of school. Everyone will say, 'gosh I can give $5!' but what you do is put them in the giving mode, and if you can work that correctly, as they have more funds, you (be it Cornell, Rochester, etc.) will benefit from their increased ability to donate money. Thanks and if I am able to help in any way, let me know."

In closing, it has been the intention of this effort to make the value of these programs and their ability to be replicated both appreciated and understood. These are but models for other institutions to consider. Each school of course needs to respond in their own unique way given their traditions and histories, communities, and individuals.

Appendix I

Traditional Work Colleges

Alice Lloyd College is located in Pippas Passes, Kentucky, in rural Knott County in the Appalachian Mountains. It has a coeducational student body of 550 on a rural campus of 175 acres. The College was founded in 1923 and is independent of specific religious affiliation. The vast majority of its students are residents of Kentucky from the immediate area and 100 percent are eligible for financial aid. All students are required to work 10 hours per week during the school year in all aspects of campus programs and services.

Blackburn College is located in Carlinsville, Illinois, a small town in southern Illinois, approximately one hour from St. Louis, Missouri. The College offers a broad-based program of liberal education and teacher preparation. Founded in 1837 and affiliated with the Presbyterian Church, the College has approximately 550 students, 75 percent of whom are residents of the state of Illinois. Approximately 90 percent of all Blackburn students are eligible for financial aid and all students participate in the College's work program. Students work 15 hours a week during the school year. Students also serve as student managers for a variety of campus programs, and a Work Committee of students makes basic managerial decisions for the entire work program.

Berea College is an independent four-year, liberal arts college founded in 1855 and located in the town of Berea, Kentucky, about 45 miles south of Lexington on the edge of the Appalachian Mountains. Its total enrollment is 1500 students with a faculty of 127. Admission preference is given to Appalachian residents with financial need. One hundred percent of Berea students are eligible for financial aid and all students participate in a labor program of ten or more hours per week, providing services for the campus, campus industries, a restaurant, and hotel.

Warren Wilson College is a liberal arts college located in Swannanoa, North Carolina, and is affiliated with the Presbyterian Church U.S.A. Founded in 1894 as the Asheville Farm School for Boys, it has 456 undergraduates from 38 states; 30 percent of them are residents of North Carolina. Of its full-time student body, 85 percent are eligible for financial aid. Its setting is a mountain valley outside the city of Asheville, on the edge of the Blue Ridge Mountains. Students are required to work 20 hours per week. Work assignments range from the College dairy to operating the power station. Each student is also expected to fulfill a 60-hour community service project as a graduation requirement.

The School of the Ozarks is located in Point Lookout, Missouri, a small community in the Ozark Mountains. The independent four-year college was founded in 1906 and currently has a student body of 1300 undergraduates and a faculty of 93. Ninety-nine percent of the undergraduate students are eligible for financial aid and 70 percent are from the state of Missouri. Students must work 20 hours per week in various campus work assignments.

Source: "Labor, Learning, and Service in Five American Colleges," *A Report to the Ford Foundation* (June 1989).

Appendix II

Methodology

In the summer of 1993, 5070[1] surveys were sent to alumni of Cornell, Duke, Rochester, and 17 of the 20 Dana schools who participated in the SAEQ program. (Barnard, Lafayette, and Oberlin participated in the institutional survey but chose not to involve their alumni in the participant research.) The mailing consisted of three waves. Wave 1 included a letter from the participating institution to introduce the research effort, including a copy of the survey itself. Wave 2 contained a postcard sent 20 days after Wave 1 to either remind or thank participants. Wave 3 was a second survey sent to all non-respondents 14 days after the postcard with another cover letter from the investigators. This third wave included those who did not list their name and were therefore unknown, as well as true non-respondents. This was done to guarantee anonymity to those who wanted more than confidentiality but did create the problem of duplicate mailings and therefore the potential of duplicate responses that had to be dealt with. The research enjoyed over a 53 percent response rate with 2797 surveys returned of which 2694 were usable. Seven institutions conducted their own mailing. The remaining 13 were mailed by the University of Rochester research team. Bryn Mawr enjoyed the highest response rate, above 73 percent; Parsons/New School the lowest at 24 percent. (See Table II-1.) Seven Dana-funded SAEQ schools chose not to participate in either the institutional survey or the alumni survey. Those schools were Cooper

---

[1]It is not possible to determine the total number of unduplicated alumni of these programs. That is to say, for example, through 1993 Cornell University had graduated 2097 unduplicated Tradition participants of whom it had addresses of 1750. Thus, the 5070 alumni of these programs who were mailed surveys likely represent 80 percent to 85 percent of all alumni in these programs if the Cornell distribution is representative.

Table II-1.

University	% Ret.	# Sent	# Ret.
Birmingham-Southern	61.68%	167	103
Bryn Mawr	71.08%	204	145
Canisius	56.71%	164	93
Dickinson	58.42%	101	59
Furman	56.55%	168	95
Parsons/New School	23.81%	63	15
Smith	64.80%	179	116
Vassar	47.62%	294	140
Wesleyan	66.67%	114	76
Wheaton	51.75%	114	59
Total We Mail	**57.46%**	**1568**	**901**
Bates	38.89%	36	14
Bucknell	63.00%	100	63
Davidson	38.82%	85	33
DePauw	50.53%	190	96
Franklin & Marshall	39.09%	220	86
Ithaca	66.67%	192	128
Union	35.50%	307	208
Total They Mail	**39.91%**	**1130**	**528**
Cornell	54.06%	1750	946
Duke	55.45%	330	183
Rochester	46.58%	292	136
Total Non-Dana	**53.33%**	**2372**	**1265**
Total Overall	**53.14%**	**5070**	**2694 received & recorded**

Union, Kalamazoo, Mt. Holyoke, Muhlenberg, Pratt, Swarthmore,and Wellesley.

Regarding the results of differences, all data reported utilized the Scheffe's Test for Significance at the .05 level. Scheffe is the most conservative estimator and therefore one can be more confident that the effects found are real. The Scheffe Test for Significance involves pair-wise comparisons of means and requires larger differences between means for significance than other multiple comparison tests. In addition, the Chi-square Automatic Interactive Detector program was utilized (CHAID). CHAID performs segmentation modeling, a relatively new statistical application. CHAID divides a population into two or more distinct groups based on categories of the most powerful predictors of a dependent variable. It then splits these groups into smaller subgroups based on other significant predictor variables. This splitting process continues until no more statistically significant predictors can be found. Not only statistically significant groups are reported by CHAID. The groups that CHAID derives are mutually exclusive and exhaustive—the groups do not overlap and each case is contained in exactly one group. Since segments or groups are defined by combinations of predictor variables, each case can be easily classified into its statistically appropriate segment by simply identifying the categories of the predictors.

Appendix III

Overview of the Results and Profile of Respondents

Of the 2694 respondents, 63 percent were female, 37 percent male, 85 percent Caucasian, 6 percent Asian american, 4 percent African American, 3 percent Hispanic American, and less than 1 percent Native American. (Table III-1 shows the gender, underrepresented minority, and non-traditional age distribution of program participants by school. The predominance of women respondents is consistent with the predominance of women participants in these types of programs; however, it is not representative of the gender distribution at these institutions. Higher responses to surveys from females is routinely observed in these types of research efforts. Table III-1 also shows a break-down of gender enrollment at the participating institutions.)[1]

In fact, two-thirds of Rochester and Dana respondents were female compared to just over one-half of the Cornell and Duke respondents. Ninety-seven percent were citizens of the United States. Eighty-eight percent were enrolled as freshmen, 12 percent transfers. Eighteen percent of The Cornell Tradition respondents were transfer students. Nine percent of the Dana participants; 8 percent of the Rochester alumni; and only 1 percent of the Duke alumni were transfers upon enrollment. Ninety-six percent were traditional students, 4 percent non traditional. Cornell and Dana respondents were significantly different from Duke and Rochester regarding non-traditional age students. Five percent of all Cornell students were non-traditional age; 4 percent Dana; and only 2 percent of Duke; and none of the Rochester respondents. Fifteen percent of the participants had a freshman year experience and may or may not have participated in subsequent years.

[1] *Peterson's School Guide,* 1994

Table III-1: Population by Institution and Total Population

College Name	Enrollment*	% Female* at Institution	% Female Responded to Survey	Institution total* Under-represented Minority	%Minority Respondents to Survey	Non-traditional Age* 25 or Older	Non-traditional Age By Respondent
Bates	1515	50%	57%	4%	0%	1%	0%
Birmingham-Southern	1763	57%	64%	13%	4%	13%	2%
Bryn Mawr	1847	100%	100%	10%	6%	4%	.7%
Bucknell	3603	46%	51%	6%	9%	1%	3%
Canisius	4859	45%	62%	8%	0%	17%	4%
Cornell	12,841	46%	56%	12%	8%	2%	5%
Davidson	1550	47%	52%	5%	10%	0%	3%
DePauw	2058	55%	55%	11%	5%	1%	1%
Dickinson	2047	56%	51%	4%	2%	0%	0%
Duke	6130	46%	56%	13%	6%	2%	2%
Franklin & Marshall	1800	45%	51%	6%	1%	1%	2%
Furman	2936	54%	57%	4%	2%	1%	2%
Ithaca	6058	54%	72%	5%	.8%	1%	5%
Parsons/New School for Social Research	6050	56%	73%	8%	31%	12%	27%
Smith	2879	100%	100%	7%	4%	11%	25%
Union	2309	44%	61%	6%	1%	—	3%
Rochester	5188	48%	66%	10%	7%	—	0%
Vassar	2272	58%	59%	14%	14%	4%	1%
Wesleyan	3331	50%	62%	16%	11%	1%	1%
Wheaton	1319	65%	98%	5%	4%	1%	0%

Source: *Peterson's School Guide*, 1994

Eighteen percent began participation in the program's sophomore year. Two-thirds of the program participants were involved in their junior and senior years only. A breakdown by year of participation and time of program (academic year vs. summer) of those who responded may be seen in Table III-2.

Table III-2.[2]

	Academic Year				Summer			
	1st	**2nd**	**3rd**	**4th**	**1st**	**2nd**	**3rd**	**4th**
Cornell	31%	43%	65%	71%	3%	9%	18%	26%
Dana	6%	15%	26%	35%	0%	8%	24%	40%
Rochester	2%	6%	12%	18%	0%	12%	39%	69%
Duke	1%	1%	5%	1%	0%	9%	42%	60%

[2]With the exception of second summer, all other participation distributions are statistically significant.

Bibliography

Abramson, Rudy, "From the Beginning Berea Nurtured Those Most in Need."

"An Invitation for Inspiration," Brochure, Berry College.

"Aspire to Something Better," Brochure, Berry College, Oct. '91.

"Berry College Student Work Handbook for Students and Supervisors."

Berry College, "1991-1992 President's Report."

Biddle, William W., *Growth Toward Freedom*, Harpers and Povothers, New York, '57.

Breiseth, Christopher N. "Learning to Hear the Voice of the Desert," *Change*, v15, n6, Sept. '83,

Caccese, Arthur "The Kalamazoo Plan: Institutionalizing Internships for Liberal Arts Students," *Innovative Higher Education*, v9, n1, Fall-Winter '84, pp. 59-74.

Carnegie Foundation for the Advancement of Teaching, "The Price of College Shaping Students' Choices," *Change*, May-June 1986.

Carter, Lindy Keane, "Charting a New Course of Study," *Currents*, v15, n6, June '89, pp. 6-13.

Christoffel, Pamela, "Working Your Way Through College: A New Look at an Old Idea," *College Entrance Examination Board*, Washington, D.C., Oct. '85, 43 pp.

Davis, James, et. al., "A Study of the Internship Experience," *Journal of Experiential Education*, v10, n2, Summer '87, pp. 22-24.

Enteman, Willard F., et.al., "Improving Undergraduate Education with Value-Added Assessment," June '86, 13 pp.

Erahn, Susan, et. al., "Student Financial Aid and Educational Outcomes: Is There a Difference Between Grants and Loans?", Association for the Study of Higher Education, Feb. '87, 31 pp.

Fletcher, Joyce K., "Field Experience and Cooperative Education: Similarities and Differences," *Journal of Cooperative Education*, v27, n2, Winter '91, pp. 46-54.

Fordyce, Hugh, "The Role of the College Institutional Research Office," *Research Report,* v13, n1, Nov. '89, 7pp.

Goldberg, Vicki, "The Soup-Kitchen Classroom, *New York Times.*

Greene, Elizabeth, "5 Colleges to Study Benefits of Requiring Students to Hold Campus Jobs," *Chronicle of Higher Education,* v34, Sept. 2, '87, p. A86.

Hansen, Janet S., "Alternatives to Borrowing: How Colleges are Helping Students Avoid Debt," *Change,* v18, n3, May-June '86, pp. 20-26.

Hansen, W. Lee, et. al., "The Impact of Student Earnings in Offsetting 'Unmet Need," Program Report 85-89," Wisconsin Center for Education Research, Oct. '85, 39 pp.

Healy, Charles C., "Relation of Career Attitudes to Age and Career Progress during College," *Journal of Counseling Psychology,* v32, n2, Apr. '85, pp. 239-44.

Henry, Nicholas, "Are Internships Worthwhile?" *Public Administration Review,* May-June '79.

"Improving Financial Aid Services for Adults: A Program Guide," College Entrance Examination Board, NY, NY, '83, 132 pp.

Kaiser, Marvin A., "Kansas State University Community Service Program," Kansas State Univ., Manhattan, Sept. '89, 26 pp.

Kane, Stephen T., Healy, Charles C., Henson, J., "College Students and their Part-time Jobs: Job Congruency, Satisfaction, and Quality," *Journal of Employment Counseling,* v29, Sept. '92.

Kisler, Christina, "Community Service Program, Westmont College," Westmont College, Nov. '89, 52 pp.

"Labor, Learning, and Service in Five American Colleges," A Report to the Ford Foundation, June '89.

Lonati, Stacy, "Internships Invaluable, Students Say; Critics Charge Statistics Exploit Free Labor," *St. Louis Journal Review,* Nov. '93.

Luzzo, Darrell Anthony, "Social Class and Ethnic Differences in College Students' Career Maturity: A Quantitative and Qualitative Analysis," April '91, 39 pp.

McCartan, Anne-Marie, "Students Who Work: Are They Paying Too High a Price?" *Change,* v20, n5, Sept.-Oct. '88, pp. 11-20.

McShane, Elizabeth, "Berea College: Carrying on Traditions," *AGB Reports,* v29, n4, July-Aug. '87, pp. 22-23.

Minges, Robert J., McGill, Lawrence T., and Schaffer, James M., "Innovations and Options: How Colleges Cope with Reductions in Federal Aid for Students," *AAHE Bulletin,* Nov. '86.

Mitchell, Catherine C. and Schnuder, C. Joan, "Public Relations for Appalachia: *Berea's Mountain Life and Work,"* Journalism Quarterly.

Mosser, John W., "Field Experience as a Method of Enhancing Student Learning and Cognitive Development in the Liberal Arts," University of Michigan, Fall '89.

"Naturally . . . Berry College," Brochure, Berry College.

Nelson, Kendra Kaye, "How Relevant Career Experiences Influence Career Decision Making," *Master of Education Report,* Aug. '90, 62 pp.

Nevill, Dorothy D., "Career Maturity and Commitment to Work in University Students," *Journal of Vocational Behavior,* v32, n2, Apr. '88, pp. 139-151.

Newman, Frank, "Higher Education and the American Renaissance: A Carnegie Foundation Special Report," Carnegie Foundation for the Advancement of Teaching, '85.

O'Brien, Eileen M., "Outside the Classroom: Students as Employees, Volunteers and Interns," *ACE Research Briefs,* vol. 4, No. 1, 1993.

Pascarella, E.T., and Steven, JR (1985), "The Influence of On Campus Work in Science on Science Career Choices During College: A Causal Modeling Approach," The Review of Higher Education, pp. 229-245.

Ramsay, William R. *America's Work Colleges: Linking Work, Learning and Service.* Unpublished manuscript April 1992.

Ramsay, William R., "Reflection on Labor, Learning and Service: A Service-Learning, Philosophy and Style in Five American Colleges," *Journal of Student Employment,* v1, n2, Summer 1989.

"Report on Student Work Programs Leadership Conference," Mount Berry, GA: Berry College, '82.

Richards, Ellen W., "Undergraduate Preparation and Early Career Outcomes: A Study of Recent College Graduates," *Journal of Vocational Behavior,* v24, n3, June '84, pp. 279-304.

Rock, Maxine, "Report on Student Work Programs Leadership Conference at Berry College," Berry College, Funded by Charles Stewart Mott Foundation, '82, 58 pp.

Rosenbaum, Allan, "Public Service Internships and Education in Public Affairs: Administrative Issues and Problems," National Society for Internships and Experiential Education, Apr. '76, 52 pp.

Rosenthal, Jonas O., "Alternative Higher Education: *The Journal of Nontraditional Studies,* v6, n4, Summer '82, pp. 214-218.

Scannell, James, "Crisis in the Work Force, Crisis on Campus, " *The Admissions Strategist,* The College Board, Spring 1993.

Scannell, James, "The Effect of Financial Aid Policies on Admission and Enrollment," The College Board, 1992.

Shane, Ruth S., "Case Study: Experiential Learning at Boston University," *New Directions for Experiential Learning,* n20, June '83, pp. 75-81.

"Small Comprehensive Colleges: Best of Both Worlds," *US News and World Report,* Oct. 10, '88.

Stampen, Jacob O., et. al., "The Impact of Student Earnings in Offsetting 'Unmet Need,'" *Economics of Education Review,* v7, n1, '88, pp. 113-126.

Stephenson, Stanley P., Jr., "Work in College and Subsequent Wage Rates," *Research in Higher Education,* v17, n2, '82, pp. 165-178.

"Student Aid Success Stories: Celebrating 25 Years of the Higher Education Act," National Association of Student Financial id Administrators, July '90, 32 pp.

Taylor, M. Susan, "Effects of College Internships on Individual Participants," *Journal of Applied Psychology,* v73, n3, '88, pp. 393-401.

"The Labor Program," Berea College internal document.

Thomas, Debbie G., "College Students Perception of the Effects of Working While Attending School," Journal of Student Employment, v2, n1, Winter '90.

Topolnicki, Denise M., "When Financial Aid is Not Enough," *Money,* v15, n9, Sept. '86.

Van der Vorm, Patricia, et. al., "The American University," *Educational Record,* v65, n4, Fall '84, pp. 60-62.

Van der Water, Gordon, et. al., "Working While Studying: Does It Matter? An Examination of the Washington State Work Study Program," AVA, Inc., May '87, 138 pp.

Weston, William D., "Competence, Autonomy, and Purpose: The Contribution of Cooperative Education," *Journal of Cooperative Education,* v19, n2, '83.

Wilburn, Rusty, "A Student Guide to Berea's Labor Program."

Index

DATE DUE

JUL 14 '98			
ILL			
966674			
3/29/98			